For Rick my rock and the world's longest serving lieutenant.

THE ENEMY DEAD

The dead are always searched,
It's not a man, the blood-soaked
Mess of rice and flesh and bones
Whose pockets you flip open;
And those belongings are only
The counterpart to scattered ball
Or the abandoned rifle.

Yet later the man lives
His postcard of a light blue
Donkey and minarets
Reveals a man at last.
'Object-the panther mountains!
Two-a tired soldier of Kiku!
Three-my sister the bamboo sigh!

Then again the man dies. and only what he has seen
And felt, loved and feared
Stays as a hill, a soldier, a girl;
Are printed in the skeleton
Whose white bones divide and float away
Like nervous birds in the sky.

Bernard Gutteridge

Military Vignettes

A Selection of Australian War Stories

Carol Rosenhain

ECHO BOOKS

First Published in 2021 by Echo Books
Echo Books is an imprint of Superscript Publishing Pty Ltd, ABN 76 644 812 395
Registered Office: Suite 401, 140 Bourke St, Melbourne, VIC, 3000
www.echobooks.com.au
Copyright ©Carol Rosenhain
Creator: Rosenhain, Carol.
Title: Military Vignettes: a Selection of Australian War Stories
/Carol Rosenhain
ISBN: 978-0-6488546-3-0 (paperback)

A catalogue record for this book is available from the National Library of Australia

Book layout and design by Peter Gamble, Canberra.
Set in Garamond Premier Pro Display, 12/17 and/Trajan Pro3 and Lucida Handwriting.
www.echobooks.com.au

Contents

Introduction	vii
Sleight Of Hand	3
A Little Tipple	15
Bureaucratic Minefield	23
A French Letter	35
So Close to Home	43
Just Reminiscing	55
A Country Cenotaph	65
If Only She Knew	75
Liberty Hall	87
The Blooding	99
Never Forgotten	109
About the Author	119

INTRODUCTION

The basis of the following anthology of short stories has been largely adopted from the files of a department called Base Records. This department was established in October 1914 just after the outbreak of World War 1. Its task was to compile, collate and maintain records of every one of the 400 000 troops who enlisted for the First World War; of whom 331 000 served overseas. The Officer in Charge of this department, was Military Staff Clerk, James Malcolm Lean. Initially he had two clerks to support him, but by the end of 1918, he was the Officer in Command of a staff of over 400 and had been promoted to the rank of major.

Fundamentally, the task of the department was to honour the army's duty of care to each of its servicemen and to act as a liaison between the army administration and the families of the serving soldiers. The department, with very primitive means of communication with those at the front, maintained among other things, files on a soldier's deployment, death or wounding.

Given the information to which they had access, the department was constantly bombarded with letters from families wanting information about their loved one. All of these letters were added to the soldier's file, as well as copies of the replies from Major Lean and his staff at Base Records. All of this material has been carefully preserved for over a century and is available to the public through the National Archives Website. In accessing this material, and combining it with additional research, I have detailed the remarkable work of this previously unheralded department in a recent publication, *'The Man Who Carried the Nation's Grief'.*

In this volume of short stories, most case names and details of the protagonists have been drawn from the files and where indicated are true. In other instances, where the events are too sensitive for publication, names have been altered, although the events are accurate. Obviously we cannot know the thoughts of these long dead Australians, so in some instances I have used authorial privilege to bind elements of the story together. For most stories, a brief context has been explained. As will be obvious, some of the material has been developed from personal accounts and varied sources.

The final two stories relate to the Second World War. They reflect the experience of my father and father-in-law as members of the 2nd Australian Infantry Force.

My hope is that in reading these stories, readers might begin to grasp some understanding of the horror, humour and sacrifice that the valiant men of the First and Second AIF made for all Australians.

LEST WE FORGET

SLEIGHT OF HAND

The events of this story are true and were drawn from the Base Records files. The names have been altered in order to protect descendants of the protagonist from potential embarrassment. Whilst the majority of men served honourably, many saw their changed circumstances as an opportunity for personal aggrandisment. William Burgess was such a person.

When William Burgess enlisted in Sydney in October 1915, he felt real fear for the first time in his life. He was not a big, brawny Australian used to mixing it with his mates at the pub or the rugby. Reserved and dapper, he was happier spending his leisure time at home reading, with his mother responding to his every whim. Needless to say, the prospect of the grisly reality of war failed to excite him as it did other young men. There was nothing about William's appearance that bore even remotely a soldierly mien. His job as a postal clerk, sorting letters and weighing parcels had done little to develop his physique or complexion. He was a fair, anaemic looking young man weighing just 112lbs and barely five feet five inches tall. His work mates jokingly called him Mr. Puniverse and fell about laughing uproariously when he spoke of enlisting. They could not imagine how the quietly spoken and almost effete young bachelor could lift a .303 rifle, let alone remain standing if he fired it.

Those same men guffawed uproariously when he turned up at the office in full uniform, laden with his kit on the day of his embarkation. 'Ya won't look that spruce in the trenches mate! Have ya got a hanky in your pocket 'cos you might need to blow your own nose out there?' came the barrage of light hearted banter from his former co-workers. They continued the ridicule with the assurance that they would read the casualty lists VERY carefully from that day on. The general feeling among the group was that if the AIF needed William, there was no bloody hope of ever winning the war. Undaunted, William smiled at the chiacking while flashing the newly inscribed gold watch his mother had given him. 'Pity you married hen pecked blokes are too old to enlist as there's no way any of you suckers could score something as classy as a Rolex, Hope youse all enjoy changing the nappies while I get to travel and become a hero,' he grinned. 'I'll be seeing you chaps, some time down the track,' he smirked as he exited with a snappy salute that showcased the gold adornment on his wrist.

Initially William's talisman seemed to ensure he had the last laugh. There were no route marches and muddy trenches for him. In their wisdom, the authorities had allocated his deployment to the Postal Corps in Edwards St, London. His peace time employment was deemed a worthy skill and William was delighted to be sent on a holiday to London at the government's expense. No trenches for him. Just a safe job in Blighty at six bob a day and a gratuity to follow at the war's end.

With over 330 000 Australian troops deployed to join the British Expeditionary Forces, millions of letters from Australia passed though London on their way to battle fields in France, the Middle East and Egypt. During his orientation sessions, William was drilled about the importance of the responsibility entrusted to him. Although mail from home took two to three months to arrive in London, and then further weeks to reach the troops, his superiors stressed that the mail was essential for troop morale. It must proceed at all cost. For the diggers, so far from home, the mail provided their only connection to and comfort from the civilised world of family and fireside.

As a private, William was initially assigned to the day shift under the supervision of a sergeant. Daily bulletins were posted as to the location and drop off points for each battalion and the hand had to work as quickly as the mind in sorting the letters into the correct satchels for distribution. William's experience, established him as a trusted and speedy worker, especially when the bulging sacks of Australian mail arrived after a ship had berthed. He delighted in emptying the sacks and scanning the stream of envelopes with carefully and lovingly inscribed names pouring out in front of him. He knew from his own personal correspondence, that the news of mothers and sweethearts probably described the mundane details of family life and the entreaty to write regularly. Yet the texture of each letter between his fingers seemed like a piece of velvet, offering intrigue and a degree of mystery. Letters were a tangible commitment and they unravelled the promise of a return to normality once this wretched war was over.

In his off duty hours, William was housed at the barracks at Chatham, which although convivial soon bored him. He began using his leisure time to become a tourist and enjoy the sights that the great city had to offer. He visited the museums and famous landmarks such as Buckingham Palace and Hyde Park. A friend had recommended the art galleries, but he didn't fancy wandering among boring canvasses filled with images he didn't understand. In his nightly forays he found his way to the seedy streets of Soho where the girls on the game offered him an artistry of a different kind. William was delirious with the excitement and comfort that these women provided. Sex was a drug that he had never tasted and once that virginal boundary had been crossed, he delighted in the addiction. To his chagrin, these delights were abruptly curtailed. His efficiency and expertise saw him promoted to corporal. The responsibility that came with the two stripes placed William in charge of the night shift!

Although he wrote to his mother informing her of the good news, he secretly resented the restrictions imposed on his private pleasure. Still, there had to be other perks to the job. Then like the prostitutes, he realised they passed before him every night. Sensuous letters, assorted parcels of varying shapes and sizes. All, mysterious and indeed very alluring. A fleeting temptation soon became a seduction. If he could purloin parcels that perhaps bore gifts, exchange for money or favours on the black market was an easy possibility. Food was stringently rationed in Britain and for a price, the Brits would relish some decent Aussie delicacies. William too was also fed up with the bland diet of powdered eggs and tasteless stews. He longed for some real Australian home cooking, a moist fruitcake, jars of jam or even a tin of condensed milk.

A keen eye could identify from the shape of parcels those which bore largesse. A quicker hand could deftly remove them from the processing line and hide them in a secret satchel. As the night supervisor, such bounty was really his for the taking. The sorting clerk could be conveniently sent on an errand, while William's sleight of hand could claim the parcel for himself. He had heard that front line soldiers often lamented the absence of parcels they had been promised, but most accepted that a World War offered few certainties. With that belief and his senior rank, he was bereft of conscience and fear of discovery.

William's first forays into petty theft were not overly ambitious. Initially he stole one package a night. Each was small enough to be smuggled out under the layers of his greatcoat in the charcoal light of the cold morning dawn. Once on the street, his pace quickened as he hurried to his quarters to unwrap the alluring bundle of mystery. Every morning became a kind of mini Christmas as he snapped the string and unwrapped the heavy brown paper. He delighted in a new pipe, tobacco, cigarettes, tins of juicy Australian fruit, or on a good day, a moist fruit cake. Occasionally, when he prised the lid off a cake that had been lovingly baked and packed by a mother, he discovered some pound notes carefully wrapped in grease proof paper. With such bounty, black market goods were within his grasp. With regular windfalls, William became emboldened, and continued luck made him confident that he could sit out the war in comfort. Surely, he reflected, a little extra cash might even entice some of the girls to ply their trade at more suitable hours for his pleasure!

Being a senior postal clerk provided easy pickings and William sustained the scam for months, confident that his duplicity would never be discovered. In such cases confidence often breeds carelessness and this ultimately led to William's undoing. One night, back at the barracks, a fellow postal employee developed a blood nose that needed staunching. He had used his own supply of handkerchiefs and knocked on William's door with a request for any that William might be able to offer.

'Not a problem Sam,' responded William as he burrowed in his drawers.

'Only bought myself a new box the other day. I think there's three in here. That should do the trick,' he proffered as he passed the box to the bloodied soldier.

'That's very generous of you Corp,' replied the young man. 'I'll keep this box and fix you up with new one when we next get paid.'

'No worries mate,' smiled William.

Alas, for William there were huge worries in store. Having stopped the blood flow with one handkerchief and cleaning his face up with the other, the young clerk decided to put the remaining handkerchief in his pocket. As he drew it out of its box a letter fell to the floor. Bearing full military details, it was addressed to a Private Stephens whose battalion was stationed in France. Reading the letter Sam soon realised that the soldier's mother had packed it with her gift as a means of safe keeping. He also realised that the recipient would never receive his package. There was only one explanation as to how this gift could have found its way into the possession of Corporal William Burgess. The next morning Sam took his suspicions to the Officer in Charge of Postal Services. Consequently, a secret surveillance operation was duly implemented. Within a fortnight Corporal William Burgess' quarters were raided and he was ordered to appear before a District Court Martial and a Civil Court.

There was nothing confident or ebullient about Corporal Burgess as he was escorted to appear before three Commanding Officers. After saluting and then standing at attention, he was informed that he faced charges of theft and neglect to the prejudice and good order of military discipline. A search of his room had revealed all manner of stolen goods. Additionally, an audit and reconnaissance of his office also discovered that he had failed to despatch 37 letters addressed to diverse units in France as his duty demanded. Instead he had squirreled the letters away and allowed them to remain in the despatching office for two months. Given the privileges and responsibility that the corporal had flouted, the charges were serious and the gravity of the punishment reflected his betrayal of trust. His crime transgressed civil, as well as military law and accordingly the *London Evening News* reported the trial.

> A Corporal of the Australian Postal Service at the GPO named William Burgess has pleaded guilty today at the Guildhall to stealing postal packets containing cigarettes and other gifts addressed to soldiers at the front. It was stated that he had been suspected and had been kept under special observation. Additionally he was seen to take the packages and readdress them to a friend of his in the Australian Expeditionary Forces. He had withheld mail for despatch and in his quarters a quantity of new pipes, cigarettes and tobacco were found, which he asserted he had purchased himself. He was reduced to the ranks and sentenced to six months imprisonment at Lewes Detention Centre.

William's incarceration was a shameful and gruelling experience and he was treated as a leper by the other prisoners. Such was his misery that he was greatly relieved when 80 days of his sentence was remitted. This was largely because the Australian Army in France had suffered an enormous toll on its fighting force. Reinforcements were desperately needed. Upon his release, William was deployed to a training battalion at Perham Downs before being posted to France. Gone was the good life, the sinecure that he had forsaken. Here was the reality of the grisly war that he had long feared. Hunger, mud, rats and fatigue topped up with an overwhelming terror became William's daily lot. His fear was indeed justified, for the ferocious battle at Bullecourt was one he was ill equipped to fight. Consequently, William was killed in action within hours of the attack. That night, as members of his platoon scoured the crater riven ground for survivors, they found William's body. As was the custom before burial, his pockets were emptied and his watch and one identity tag were placed in a bag to be forwarded to London for dispatch to his next of kin in Australia. That was the theory, the principle. The practice, in so many cases was for a member of the burial party to souvenir a watch, a cigarette case or a fountain pen, which had been inscribed by a parent or partner as a loving and valuable farewell gift. By sleight of hand, William's mother's gift moved from his limp wrist to the pocket of an anonymous soldier.

Back in Australia William's mother was naturally devastated at losing her only son. There were no other children. For her, his death was the end of the penny section. All she could hope for was the return of his belongings, a shirt with which she could capture his smell, a diary recording his reflections and the gold watch which he would have worn when he fell. She needed these tangible mementos to hold onto the essence of flesh and blood that had returned to dust. Months later, she was informed by an official military letter that a parcel of William's effects had been received by Base Records in Melbourne which his mother directed to be forwarded to her home. Longingly she awaited the postman's whistle each day. Finally, she received a directive to collect the parcel from her local railway station to which she rushed like a woman possessed. In the stationmaster's office she quickly bundled the crumpled brown package under her arm, before hurrying home, to unwrap the remnants of her beloved son's life. Thundering through the battered back door and flinging the parcel on the kitchen table, she slashed the binding with manic intensity and ripped at the heavy, brown paper. With the contents unravelled, her hopes were dashed when she gazed upon her son's few paltry possessions. There lay dear William's disc, some shirts, her letters and some items of shaving kit. There was no diary of her beautiful boy's noble thoughts, but more disturbingly no trace of the gold watch that she had sacrificed so much to give. Naively, she had thought that perhaps it had stopped at the moment he had been killed, and with its return, she could hold that moment in her heart forever; but there was nothing.

Desperate enquiries to Base Records reassured her that all of the items had matched the inventory and there was no further action that could be undertaken. Her grief shuffled between anger and depression. She could find no comfort or acceptance at having so little of the boy she had nursed and nurtured. Bereft, with no family it was the neighbours who bore the brunt of Mrs Burgess' grief. One Sunday, on the anniversary of William's death, her neighbours sensitive to the occasion, accompanied her home for a cup of tea after church. The parcel which had been retrieved from a cupboard that morning still lay on the kitchen table. Seeing it and understanding a mother's pain, her guests stood silent as Mrs Burgess drew their attention to the shirt that she lovingly held to her face in a forlorn effort to inhale the last vestiges of her son. They understood that the pain of their friend's loss was almost unendurable but knew that all the will in the world would not bring him back. For Mrs Burgess, the moment was overwhelming and tears turned to sobs as she vented both her fury and frustration at never receiving the precious watch William had worn at his death. Unable to reconcile its absence from her parcel of belongings she articulated her inability to understand its absence.

'After all,' she wept, 'If they were able to return his disc, which they must have removed from his body before they buried him, why wouldn't they have removed the watch at the same time to add to his effects?'

One kindly neighbour paused before she spoke, as if wondering whether to offer a harsh truth to her grieving friend. Then mustering her resolve she whispered.

'I've heard that there is a lot of theft that goes on over there. Friends of mine who 'ave lost boys, always say that the valuables never come 'ome.'

She did not finish the sentence before Mrs Burgess dabbed her eyes and interjected incredulously.

'Surely no one would steal from a fallen soldier, a mate in the same unit! I can't and won't believe that any member of the AIF would steal from the dead or the living when we are all united in this ghastly war.'

'Well Ducks,' consoled her friend, 'Times is 'ard and folks have to graft what they can. There's many that lack the decency of your family's values. But if that belief gives you comfort, who am I to convince you otherwise?

A LITTLE TIPPLE

The First World War imposed particular hardships on members of the AIF. Too far away for home leave, they often found solace in drink or other forms of undisciplined behaviour. Their scant respect for authority regularly drew the ire of the British Officers while their impulsive behaviour drew condemnation from the High Command. Before their legendary qualities as a fighting force became enshrined, the Australians were often perceived as louts and drunkards.

As he signed his name to the proceedings of the formal Court of Inquiry, Major Corfe felt sickened by the details of the day's events. It was bad enough signing a Field Service Report for men who were shredded by machine gun bullets or blown to dust by shells, but this was almost worse.

What a tragic waste of lives these so called 'Accidental Deaths' proved to be. While the stress of combat certainly drove the men to drink in the precious few hours they had out of the line, it was shattering to think that some literally drank themselves to death. Sapper Bone had only been in France for a month and here he was, dead by his own hand. Similarly, Private Rhodes had only been in France for all of three months and he too had contributed to his own demise. Perhaps they were heavy drinkers before they enlisted and simply could not break the habit he pondered. Whatever the background, their deaths came at an absurd cost to the AIF and of course their families at home. Major Corfe determined that he would not discuss details at the mess that evening. It would just add fuel to the British Officers' beliefs that the Australians were an ill disciplined mob of drunken louts.

Bone's situation worried him greatly. He was aged 40 on enlistment and as a miner was assigned to the 3rd Tunnelling Company. According to his Attestation Form he was married with two sons. No doubt his was a noble commitment, but war was not a place for middle aged men with family responsibilities. Maybe the army provided an escape from the tedium of domestic life, just as a drink of alcohol offered respite from the grinding fear and labour involved in tunnelling. Still, Bone's case seemed hard to fathom. According to the witness statement of the Company Sergeant Major, Sapper Bone was present at the morning parade and went out with the working party. He stated categorically that he noticed no indication that Bone had been drinking. Another witness, Sapper Llewellyn of the same company, reported that he had met Bone at 6.00 p.m. in a street at Bapaume and that he was completely sober. Llewellyn's description of what transpired later was of greater concern.

Apparently the two had gone for a walk on the Cambrai Road where they met two unknown diggers with a bag containing bottles of whiskey. When offered a nip, Bone and Llewellyn had responded with alacrity. They were soon joined by others. Behind a hedge, five men shared the first bottle, six the second and about ten the third bottle. The whiskey had been disguised in short, black Delaware bottles. According to Llewellyn, after he and Bone had enjoyed a swig from each bottle, they then made their way back to their billets. Llewellyn then reported that after they had walked about a quarter of a mile, Sapper Bone started to struggle and needed assistance to the town square. He then collapsed.

Although Llewellyn called for Red Cross Assistance and Bone was taken to the 35th Field Ambulance, he did not regain consciousness and died the next day. Pursuant to King's Regulations 674 and the findings of a post-mortem, Major Corfe had no alternative but to declare that the soldier was not on duty when the mis-adventure and death occurred. Consequently, no one was to blame but the man himself. Surely, thought Major Corfe, Bone could not have consumed sufficient alcohol to render him paralytic? Yet on medical expertise and witness statements, he had no alternative but to reach the inevitable finding that the man had died of alcoholic poisoning.

The next case presented to him that afternoon was just as disturbing.

Private Rhodes had failed to present for a working party early one morning when billeted in the town of Bapaume. When his sergeant had seen him peering out of a window on the second floor of his billet, he ordered the soldier to put on his equipment and join the platoon. When Rhodes failed to appear, the sergeant investigated and found the man lying down on his bedding which was beside the window. It was clear that the soldier was intoxicated, and unresponsive. Recognising that Rhodes was incapable of any productive activity, the sergeant let him be and proceeded with the working party. He reported that he did not see Rhodes again until later that night. He was returning with his men, when he witnessed Rhodes stretched out on the cobble stones under the window of his billet. Rhodes' face and head were badly bloodied and a large bone in his leg had pierced his trousers. Recalling the morning's events, the sergeant surmised that the soldier had most likely roused himself from his stupor and fallen out of the window. The sergeant immediately organised his men to find a stretcher and carry the man to the local Field Hospital. Tragically Private Rhodes was declared dead on arrival. In the fall he had sustained a fractured skull and a broken femur.

There was no denying that both men had died as a consequence of excessive alcohol, but Major Corfe had agonised over the findings and documentation of the proceedings. He was able to declare with impunity that Rhodes had died as a consequence of an accident. Yet his hand was forced to declare Bone's death as a result of alcoholic poisoning. It galled him as he did not know the men. What comfort could he, as their Commanding Officer, offer to their next of kin in the usual notes of condolence. Perhaps both had been temperate men, unaccustomed to drink of any kind? Perhaps, the amount that they had consumed was alien to their systems which had reacted adversely? Reluctantly, he surmised that their brief exposure to this rotten war had probably driven them to drink. Major Corfe knew that the information communicated to the deceased servicemen's families would indeed be bitter tidings. It was hard enough to lose a husband, father or son, but in such ignominious circumstances, what solace could be found.

As he headed for the mess that evening Major Corfe speculated on the events that each of the men might have experienced in their brief months in France. After dinner, back in his quarters, he called for the men's service records. As a tunneller Bone had lived like a mole, working in darkness, no doubt with nerves on edge to detect counter activity from the Germans. He had worked with the British tunnellers to lay the mines beneath Messines. Those months of secret, painstaking work had resulted in the biggest explosion ever witnessed. Over 10 000 Germans were estimated to have been killed in that monumental blast which subsequently created chaos among the survivors. Given the impact of that strategy, Sapper Bone had certainly made a valiant and selfless contribution to the war effort.

As an infantryman, Private Rhodes had survived Pozières, a battle that had done much to promote the valour of the AIF and halt the German advance. Both men, just lowly privates, had sublimated their identity for the collective advancement of their respective companies. Their efforts had contributed to a brutal success, but perhaps the deeply ingrained morality of a distant civilisation on the other side of the world haunted their conscience. Sadly, Major Corfe reflected that whatever Sapper Bone and Private Rhodes had contributed to the First AIF, their comrades and their families would only remember them as drunks.

As he steadied his hand to pour a stiff whiskey, Major Corfe cursed the lot of the common soldier.

'Jesus bloody Christ,' he muttered to himself as he raised his glass in a silent toast. 'It's a bloody cruel world when a man dies in the service of his country and is nobody's hero!'

BUREAUCRATIC MINEFIELD

This story offers some insight into the way legislation discriminated against women in the early apart of the twentieth century and the injustice that was so often inflicted on separated mothers of deceased soldiers. The content of the story is true but names have been altered.

Daisy Fraser was only 18 when she married Norm Gifford at the Nathalia Methodist Church. As she walked slowly down the aisle, she felt the quickening in her womb and delighted in the fact that a fresh life would grow into the new century. She was consumed with excitement at the glorious future that lay ahead. Although her father was long dead and her mother was angry at her condition, her older brother gave her away, with as much dignity as was possible in the simple wooden church. Daisy was confident that she and her husband could support a fine family with the steady business the town blacksmith demanded.

She had met Norm Gifford when she left the family farm to find work in the town. Daisy had responded to an advertisement for a bakery assistant and had been thrilled to find her application accepted. Her older bother, Tom, had written her letter of application, as Daisy's schooling had been limited by her mother's need for her on the farm. With joy in her heart and a spring in her step, Daisy had left the drudgery of the farm and taken lodgings in Mrs Brew's boarding house. It was only two doors down from the bakery, which was convenient as she had to begin work at 4.00 a.m. every morning. Her day began with helping to knead the bread, after which she monitored the great ovens before serving in the shop until closing time at 5.00 p.m. It was a long and arduous day but she had developed both strength and stamina on the farm. She thought herself lucky to be earning her 12/- a week.

Norm Gifford was a good 10 years older than Daisy and his smithy was opposite the bakery. As a blacksmith it was natural that he was of a solid build. With a weathered face, huge shoulders and arms like pistons, he cut a formidable figure to anyone out late at night. However in Daisy's eyes he was like a God. Every day at 1 o'clock he lumbered into the little bakery to purchase his lunch. There he would sweep off his cap, run his fingers through his sweaty hair and ask in a flourishing manner if she would fill his order. She never failed to be astonished at his request. She had never seen a man eat three meat pies at one sitting, quickly followed by a cream bun. It was not difficult to understand though, that Norm's stomach demanded as much fuel as he heaped onto his forge each day.

After her first tentative weeks at the bakery, Daisy began to notice that Norm Gifford began to linger over his lunch. When he had finished eating she would hustle to remove his plate and cutlery. Initially he grunted, but this was soon replaced by a 'thank you,' and eventually remarks about the weather. These were soon followed by comments about her hair, or general bearing. She knew he was taken with her. For the first time in her life she felt flattered and thrilled that a man of his status should find her attractive. Naturally when their paths crossed at the close of day as each made their way home, a meeting was contrived in the street, from which of course developed conversation and a longing that Daisy had never known.

Life on the farm had been hard and isolated and Daisy had never cause to think of a future beyond the boundary fences. However in town, with her new independence, she not only had money to buy ribbons and combs, but the freedom to engage with all manner of people. Her growing confidence nurtured an emerging personality. Her newfound independence and freedom from the censure and the control of family gatherings, made her easy prey to Norm's sugary words and grandiose promises.

Their child, a boy, named Archibald Norman Gifford was born on the 26 January 1896. Although he was hale and hearty, his broad shoulders prolonged Daisy's labour and she remained fragile for months after the birth. It was not until some years later that she learned that she would bear no more children. Naturally, her only child Archie, became even more beloved.

Norm was devastated to learn that there was to be only one son from their marriage. He was a practical man and knew that disease and accidents in the district accounted for high childhood mortality. He wanted a proper brood to ensure his name for posterity. Furthermore, the trauma of Archie's birth brought pain rather than joy to the marital bed for Daisy. Norm was at first bewildered, then angry at the denial of his conjugal rights. In his despair and frustration he began to frequent the local pubs. He had formerly been unaccustomed to strong drink and it did not take much to make him a sour drunk and a bully. Daisy was shocked and frightened at this turn in his nature, especially when he began to take his anger out on her. She always hid Archie when she knew it was time for Norm to roll in from the pub. She would take the beatings and bruising, but always ensured that her beloved boy was out of range.

Gradually Norm began to return home later each night and the grapevine soon let Daisy suspect that he sought his favours elsewhere. Shamed, she continued to hold her head up high and devote all of her love and energy to Archie. Small towns are unforgiving places though, and as it became common knowledge that Norm was a drunken philanderer, business at the smithy began to wane. As the pounds for housekeeping fell to shillings, Daisy regained her previous position at the bakery to supplement the family income. She worked like a navvy at home to ensure that all of Archie's needs were met. Long into the night she patched threadbare clothing, cleaned and cooked in order that her beautiful boy should want for nothing. Norm neither noticed nor praised her efforts and it was not long before he failed to return home altogether. The first Daisy knew of his public disgrace, was when the smithy closed. Rumour had it that Norm had moved in with a woman, somewhere near Swan Hill.

Undaunted, Daisy stayed in Nathalia and committed herself to her son. Unlike her, he would be well schooled. She determined to work 24 hours a day if it guaranteed that her boy would know his letters and numbers. She hoped that one day he might not only support her in her old age, but earn a living as an educated man and hold his head up high in spite of no fatherly example. As it happened, Archie was a bright student and when he left school in 1910, he secured himself a position as a clerk in a local solicitor's office. He liked the job well enough, but felt that there was something missing in his life. It was beyond identification or articulation, which he could only sense as a kind of restlessness. Nonetheless, between 1910 and 1914 he dutifully worked six days a week. Meanwhile, the hard years and long days had taken their toll on Daisy. Although she kept house for the both of them, she could no longer maintain her position at the bakery. In spite of a small wage, Archie supported them both.

On this meagre income, Daisy and Archie managed to get by. She adored her tall, handsome son who was as gentle as his father was coarse. In turn he adored his tiny, frail mother, whom he knew had suffered both ignominy and hardship to ensure that he would always have the best that she could offer. He loved the tiny township and the bush beyond. After church on Sunday, he and his mother would walk awhile beside the dusty tracks, listening to the screech of the cockatoos and inhaling the heady scent of the eucalyptus trees. However, Archie had never travelled further than ten miles from the town and he often wondered about the world beyond of which he had read much. It was on these reflective Sundays, that he felt that undefined yearning and a restlessness that trembled in his body like a shiver.

Britain's declaration of war against Germany on 4 August 1914 and recruiting posters pinned to the door of the Shire Hall, soon galvanised the latent longing in Archie. Instinctively he knew that this was the opportunity for which he had long wished. As a soldier, he would get to travel to towns and countries that he had only dreamed about. He could escape the office, embrace adventure and perhaps distinguish himself with deeds of valour. The values of the great empire had to be protected and he felt keenly that it was his duty to enlist. His excitement was not shared by his mother. Although she knew little of politics, or the imperialistic designs of foreign powers, the enlistment application aroused fear and trepidation in her whole being. How could she surrender her beloved son for whom she had sacrificed so much? How would she survive without his physical and financial support?

Such concerns were overwhelmed by the patriotic surge of young men in the town. Even those in the surrounding countryside jumped at the opportunity to enlist. Reluctantly and with the promise of 4/- a day from Archie's 6/- a day wage, Daisy signed the parental approval note required for her son as he was under 21 years of age. Momentum gathered throughout small townships as young men like Archie farewelled tearful mothers and family at packed stations. Sons waved farewell blithely with promises to write regularly. As soon as he had been attested and been transferred to camp at Broadmeadows, Archie fulfilled his pledge. His first letter expressed the ebullience of youth in the company of men in uniform. The routines and disciplines of army life seemed to have transformed him from boy into man almost overnight. Among other details, he told Daisy that all of the soldiers had been directed to make a will and that she was listed not only as his next of kin, but the beneficiary of everything he owned in the unlikely event of his death. Further to this, he enclosed a studio photograph of himself in full uniform. It was a smiling and relaxed pose that enhanced his maturity and seemed to announce that No 4280 Private A.N. Gifford of the 5th Battalion A.I.F. could hardly wait for the next chapter in his life.

Daisy sobbed as she read Archie's letter. After Norm had left her, Archie had been her life force, her comfort and companion; her *raison d' etre*. All she had to sustain her until his return was the photograph which she placed in pride of place in the centre of the mantelpiece. At the closing of each day she would draw it to her bosom, kiss it and offer a prayer for his safe return. Later letters informed her that after the training at Broadmeadows, Archie had embarked on HMAT *Orvieto* which left Melbourne on 21 October. As much as she wanted to go to Melbourne for the final farewells, the train fare was beyond her means. She had to console herself with the photograph.

Daisy was later surprised to receive a letter from Archie with an Egyptian stamp. As far as she knew he had been bound for Europe. However she rejoiced in the news of his good health and delight in the amazing sights of the pyramids and the exotic wonders of the Middle East. Her last communication from him dated April 1915, was full of excitement as he anticipated that his battalion would soon be moved to the front. When she read that 'the boys were longing for a scrap,' her bowels turned to liquid as foreboding engulfed her. She watched the daily papers and by 30 April they were reporting the baptism of fire of the A.I.F in an invasion in the Dardanelles. She had no idea where this place was, but the mention of casualties exacerbated her fear.

Over the following weeks news of the landing and limited casualty lists filtered through to the press. Archie's name was not listed, but word had circulated that clergymen's visits to local families, and telegrams from an Army Department called Base Records, had brought dreadful news to many in the district. With no correspondence from Archie, Daisy decided to make her own enquiries to this department. The newspapers had made clear to families the correct protocol to communicate with Base Records. Tentative, though determined, Daisy wrote to Captain Lean asking if he had heard any news of her son, No.4280 Private A.N. Gifford of the 5th Battalion.

Captain Lean's initial reply was reassuring. No information had been received about her son and in such circumstances he advised her to assume that he was with his battalion. The reply offered some comfort to Daisy, but as May moved into June she wondered why Archie had not written. Living on hope it came as a shock to her to receive a telegram to the effect that her son had been wounded. Again she wrote to Base Records, this time seeking details as to his condition and whereabouts. A prompt reply followed, assuring her that the relevant information would be despatched to her immediately such details came to hand. What could she do but wait?

As August dawned, like thousands of mothers across the land, Daisy read with trepidation, reports of the landing which were now starting to emerge. It had been a fearsome battle and casualties continued to be listed. Yet again she drew on her courage to write to Base Records. Surely, she felt, they must know something of her son by now! They must know what hospital he was in. When the reply arrived she could barely slit the envelope for the tremors that engulfed her whole body. Her fear proved well justified. The letter informed her that her son was now reported wounded and missing, but endeavoured to reassure her that as next of kin, she would be kept informed of any further details. Panic consumed her. What could she do? To whom could she turn? There was no one. She just had to wait and hope and pray.

In the months that followed, Daisy daily caressed the photograph of her son. In spite of frequent letters to Base Records, no news was forthcoming. Archie was not listed in any of the military hospitals, no witness could recall seeing him after the landing at Gallipoli and he had not signed for any pay since early April. It was as if No. 4280 Private A.N Gifford 5th Battalion had not existed. With news of the withdrawal from the Peninsula at Christmas, Daisy had begun to believe the inevitable, but with no official confirmation from the military authorities that her son was dead, she clung to the frailest of hopes. Having not expected as many casualties at the landing, and with so many men unaccounted for, the military powers were struggling to determine the fate of individual soldiers. Courts of Enquiry were convened in Egypt and on 20 May 1916, Private Gifford was officially declared killed in action, burial place unknown. On the death certificate forwarded by Base Records, the date of his death was determined to be the 25 April 1915.

Daisy was devastated. After an excruciating wait of 13 months, her son was finally declared dead. Both grief and anger consumed her. Why had the news taken so long to reach her? Why were there no details of his remains and a burial? Was he killed instantly or did he suffer? No one could answer her questions. How could the military authorities be so cavalier about the loss of so many beautiful young men whose lives promised so much for the future of the young nation? She knew of neighbouring families who had lost sons and husbands, yet in most cases they could lament their loss with extended family members. For Daisy, Archie's death seemed like the end of her world. Over the next four years Daisy's health deteriorated, as without Archie's support it was difficult to make ends meet. His gratuity money and deferred pay had been minimal, as the duration of his service was short. None of his personal effects were found or returned. No details of his burial or final resting place could be identified. She would have given her life if it would have saved her son. She longed to know his fate, but all she had of him were crumbling letters and the increasingly fragile photograph.

In June 1920 Daisy was surprised to see again the once familiar brown envelope of Base Records Melbourne in her letterbox. For a brief moment she hoped that it was news that the Commonwealth War Graves Commission, of which she had read, had recovered her son's remains. Such was not the case. The letter, inquired:

Dear Madam,

It is noted that you are registered on the records of the late No.4280 Private A. N. Gifford as next of kin and so that the instructions under "The Deceased Soldiers' Estates Act 1918" may be properly complied with when disposing of war medals etc. I shall be glad to learn whether there are any nearer blood relations than yourself to the above named. For instance, is his father still alive, if so I shall be much obliged for his name and address at your earliest convenience. The provisions of a will have no bearing on the distribution of medals unless they are specifically mentioned therein, such mementos being handed over in the following order of relationship, unless good and sufficient reason for varying the procedure are stated.

Widow, eldest surviving son, eldest surviving daughter, father, mother, eldest surviving brother, eldest surviving sister.

Thanking you in anticipation of an early reply.

Yours faithfully

J.M.Lean.

It took Daisy numerous readings and almost a week to comprehend and digest the focus of the correspondence. As an understanding filtered through, she felt broken yet again. She had so little of her son. Others in the district had already received their son's medals and Daisy longed for

Archie's, to honour him on the mantelpiece beside his photo. How could the authorities determine that Norm was a more worthy recipient of the mementos of her son's service that she? Daisy ruminated on the fact for days. Nonetheless, fearful of authority and knowing full well the limitations of her own literacy, she painstakingly composed a letter to the now, Major Lean, the Officer In Charge of Base Records

> Dear Sir,
>
> You wrote some time ago asking if my husband were alive and if so to send his address as the father is consider the next of kin and what is a mother consider. She brings the child to the world and watches and cares over it all its life and has the anxiety and worry rearing it. Who has worried more over my poor son than I. He was all I had in this world as he was my main support. I have lost my health since he were killed as the shock took the use of my legs. Surely god will give woman a better place when she leaves this world and as we are treated as nothing on earth, we will be treated as something in heaven.
>
> You have never gave me any help in any way as I gave all I had for king and country. No one no how I miss him and I long for a meadle to treasure. My husband Mr Norm Gifford can be reached C/O Post Office Swan Hill. He will not value anything like it as he did not see Archie for 10 years before he enlisted. I hope you send me the meadle as I loved my son so much and god no how much I miss him.
>
> Yours truly
> Daisy Gifford.

As she blotted the letter and put the lid on the ink bottle, Daisy offered a prayer for justice. It was not to be forthcoming. She who had given all had lost all. The determination of the Deceased Soldiers' Estates Act of 1918, reflected the values of a world of men and a paternalistic hierarchy. Men had called her son to war. Men had killed him and now an unworthy man was to be lauded for his sacrifice. With bitter tears coursing her quivering cheeks, Daisy began to understood that women's rights were as elusive as quicksilver. With trembling hands she reached for the crumpled photograph and drew it to her breast. Its once brilliant definition had now faded. The light was lost, never to be illuminated again.

A FRENCH LETTER

This story was written in response to a letter that Major Lean as Officer in Charge of Base Records received from a young French woman. The letter was written in French and he took the trouble to have it translated so that he could respond to the best of his abilities. The circumstances were not uncommon, given that surviving members of the 1ˢᵗ AIF had been away from home for over four years. Names and dates in this story remain true.

There was little that Major James Malcolm Lean did not know about the exploits of the men of the First AIF. As Officer In Charge of Base Records, Melbourne, he was privy to both the virtuous and nefarious deeds of the Australians at war on the other side of the world. While Battalion Commanders informed him of deaths and deeds sustained in battle, his name was currency in most military circles as the font of all information for families and friends making enquiries about serving militia. By and large this unsolicited correspondence came from mothers or sweethearts requesting additional details about the fate or whereabouts of a loved one. He was accustomed to that. However invariably correspondence was directed to him from most unexpected sources which drew heavily upon both his diplomacy and extended networks.

Such was the case one day in August 1919 when he opened a letter addressed in French which had miraculously reached his office. The stamp was smudged and the hand writing bore a distinctly feminine curvature. His interest piqued, he carefully slit the brown envelope to see a neatly written communiqué in French. Although a man of considerable linguistic dexterity, the French was beyond him, but he could discern that the correspondent was a female, signing herself Isobelle Cisseraud.

Whether a letter was written in Swahili or Hieroglyphics, it was Major

Lean's duty to identify the concern and then address it. Undaunted, he immediately despatched a clerk to the Foreign Languages Department of the University of Melbourne to receive a translation. Although the translation confirmed his initial suspicions, his natural compassion grieved him sorely as he read the young woman's plea.

> Dear Sir,
>
> I beg to bring to your knowledge that having been engaged to be married during five months to soldier George E Sharp 5880 AIF 18 Battalion during his stay at the convalescent camp Eault Equition, this man was made welcome at my home and promised to marry me. When at the end of January 1918 he left to rejoin his unit at the Havre leaving me in the family way I informed him of that fact. He answered several of my letters and the last he wrote was dated 17th March 1918 telling me he was sorry to know the state I was in but that he hoped soon to get leave so that we could talk the matter over whilst telling me at the same time he was returning to the trenches. I therefore held these letters in reserve and since then I have heard nothing from him. I am therefore interested in his fate. I know from enquiries that I made in July 1918 that he was still with the British Expeditionary Forces and I continued writing to him without getting any reply. When my child was born on the 3rd October 1918 I informed him of it by registered letter which he accepted since the receipt with his signature on it has come back to me.
>
> I therefore seek recourse to your goodness and kindness as an Officer to do what should be done in such a case and let me know the result. I am an orphan, no mother or father and without means of supporting my child.
>
> Please accept the assurance of my respect.
>
> Isobelle Cisseraud

Although a man of Victorian rectitude, Major Lean was sufficiently worldly and pragmatic to know that this was not an isolated case. After all, some of the men had been away from Australia for more than four years and it was natural that they would yearn for and establish intimate relationships with the attractive mademoiselles. Nonetheless the poignancy and desperation of this letter touched him deeply. He doubted that he could be of assistance, but determined to unearth further details about

Private George Sharp. His own moral values wanted to make the soldier accountable to this poor young woman and then again, the reputation of the First Australian Infantry Forces was at stake too.

Even though the conflict had been over for nearly a year, Base Records still maintained records of all who had served in the Great War. There was still much to be done in repatriating soldiers and distributing medals and mementos to those who served. Delving into his meticulously maintained indexing and filing system, Major Lean was able to establish that George Edward Sharp had enlisted at Dubbo, NSW on 25 April 1916. At the time, he was a 30 year old widower with a ten year old son. His wife had died in 1913 and perhaps a combination of grief, duty and the rigours of sole parenting had encouraged him to seek new horizons. Given that he already had a son who had presumably been cared for by relatives, Major Lean could appreciate acknowledging an illegitimate child in France was a delicate and sensitive issue. It was a serious concern though and one that he decided to refer to the Assistant Adjutant General of the 2nd Military District.

Confronted with the details, the Commanding Officer was not impressed. George Edward Sharp was duly requested to attend the Paddington Barracks Sydney to clarify and discuss Isobelle Cisseraud's situation. It proved to be an unproductive interview in terms of verification and accountability. The former Private Sharp declared that he had been interviewed on three occasions by Major Robertson of the 18th Battalion in France regarding the matter. He insisted that he knew nothing of the claims laid against him and stated that he did not sign any receipt or communicate with Isobelle Cisseraud in any way. The Assistant Adjutant General was neither convinced nor amused.

'Mr. Sharp, you seem to treat this very serious matter in a most cavalier fashion. It is not a joke and whilst you may lack integrity and honour, the Australian Army has a fine reputation to uphold. You must accept responsibility for behaviour that is clearly yours!'

'I accept nothing and absolutely refuse to take any action whatsoever.

You rightly addressed me as Mr. Sharp when I entered your office and that is who I am now. I was demobbed in April, this year of our Lord 1919 and so Sir, am no longer a member of the armed forces and therefore, Sir, my life no longer comes under your jurisdiction. So Sir, you can bugger off. I wash my hands of this whole affair. Get stuffed!

With that George Sharp insouciantly headed for the door which he delighted in slamming behind him.

The Assistant Adjutant General thumped his fist on the desk and fumed. The bastard was right. Neither he nor the Australian Army held any authority over George Edward Sharp. With both bitterness and frustration he dictated a letter to Major Lean describing the outcome of the most unsatisfactory interview.

Upon receipt of this correspondence Major Lean was most distressed. In desperation he trawled through every conceivable document he could lay his hands on about Mr George Edward Sharp. Sadly his search only confirmed his initial contempt for the man. Having married the 17 year old Elsie Alliston in 1905, his treatment of her was so appalling that she retreated to her mother's house to escape his philandering and the beatings she received when she complained. Unable to tolerate his cruelty she applied for a divorce in 1906 with the *decree nisi* granted six months later. It would seem that Mr George Sharp had form.

Obviously Sharp was correct. The Australian Army no longer held any redress for his behaviour, nor could it do anything about his past. The additional details unearthed by Major Lean though heightened his compassion for the destitute young French woman. Duty and protocol demanded a reply to her correspondence. Major Lean would see to that. It was with a heavy heart that he informed the young mother that as George Edward Sharp was no longer a member of the AIF, no action could be taken by his department. Even though he knew she could ill afford it, he gently counselled that if she wished to proceed further, she should put the matter in the hands of a solicitor.

Of course poverty and the ravages of war upon the French citizens and their land ensured that George Edward Sharp would take his secret lies to the grave in 1969. True to form, he remarried in the 1920s, which again ended in an acrimonious divorce a decade later. Devoid of care or responsibility, he appears to have careened through life wreaking grief and havoc at every opportunity. Nonetheless, he would never have imagined that his unconscionable behaviour would remain archived for over a century. George Edward Sharp's French born son gave him grand children who no doubt had children of their own. With the digitisation of the whole of the Base Records archive and its global access, perhaps one day a young French traveller may knock on an Australian door to legitimately claim his birthright.

SO CLOSE TO HOME

With communication confined to brief telegrams and simple hand written letters, families often never knew the real fate of their loved ones. This story explores such a situation.
The names and events are true.

'Mum,' yelled Alan in alarm as he thrust the half folded newspaper onto the dining room table. 'Mum. Have you seen this morning's *Telegraph?*' he questioned urgently.

'No love,' responded his mother. 'What's such urgent news then?' she enquired as she bustled into the room whilst drying her hands on her apron. Thrusting the newspaper towards his mother with a frown, Alan stuttered, 'It says here that Arthur's dead, look for yourself! His name is in the casualty lists for the day before yesterday, the 9th of December!'

'Jesus, Mary and Joseph,' exhorted his mother as she clutched at the crumpled paper. 'It couldn't be him, he's on his way to England. Is there a full name, unit and number listed?'

'See for yourself mum. Sergeant Arthur Charles Percy Thwaites A.A.M.C.'

'I do see it son, but I don't believe it. There's a mistake been made. The military don't allow names to be published before the family is notified about any casualty. I know from Mrs Bourne on the corner of Church Street and the Parramatta Road. When her Ted was killed at Pozières, a clergyman came with a telegram for her. That alone was a helluva shock, but it was a fortnight after before the full particulars was published in the paper.'

'Be that as it may, all the details are correct. Where there's smoke there's fire.' Alan retorted, stabbing the familiar name with his forefinger.

'Don't think so negative son. Our Arthur is safe on a ship sailing for England and now that the Armistice has been signed, there's nothing to harm him. It's a printin' error.' She added hopefully.

'I dunno mum. I reckon I might just walk down to the barracks at Paddington and see whether they know anything about it. The army doesn't make those sort of mistakes. All information has to be checked, probably three times over before they release names to the press, and even then they have to obtain the family's permission. I'll just have to be late for work. This is more important'

Grabbing his hat and straightening his tie, Alan strode to the front door and was gone before Hannah Thwaites could object. The slam of the door seemed to reverberate throughout her tiny body and a tremble forced her legs to give way as she lunged for a chair with a fear that she had initially repressed.

'It must be a mistake,' she repeated to herself in an effort to deny any possibility of validity. While Alan was always one to go off half cocked, Hannah, ever the pragmatist, placed her chin on her cupped hands and rationalised why the details could not apply to her son Arthur. She reflected how he had always been quiet and studious and how his qualifications as a Dispensing Assistant at the Mudgee Pharmacy had kept him on the home front for most of the war. She had been well pleased that his employment was a restricted occupation, but finally the white feather brigade riled his conscience. He was 29 years of age when he enlisted in December of 1917. His qualifications were immediately valued and she was quietly relieved that he'd been posted to hospitals in Liverpool and the Garrison in Sydney for most of 1918. Then in October, he was assigned to accompany a thousand troops to join the expeditionary forces in France. She recalled Arthur's last letter. He had said that he was leaving from Adelaide on HMAT SS *Boonah*. She didn't mind that, but was greatly relieved to learn of the Armistice in December. She then knew Arthur would never see action, but she did wonder where the ship and the troops would end up. Nonetheless, she was pleased that he had been deployed as no one could justifiably call him a shirker. Hannah must have sat there mulling over all of these personal details for hours before she heard the door slam with Alan's blustering return.

'Well they know bugger all at the Barracks mum. I asked for clarification about the newspaper name but they said they knew nothing. Reckoned if they knew nothing, then in all likelihood Arthur was with his unit. They couldn't explain how his name was listed and said that details like that were compiled at Base Records in Melbourne. They promised that they would send a telegram to that lot seeking verification or denial. Can't say I'm impressed or convinced. The truth has always been hidden from the likes of us in this bloody war. Now that it is over, and just when we thought that Arthur was safe, his name mysteriously appears in the casualty lists. I don't like it mum. There's more to this than meets the eye. I'd like to think that the whole business was another bloody army balls up, but it's put the wind up me mum. I don't like it. Now that the war's over, what do they care for one bloke who cops shit at the end of it,' he spat with contempt.

'Now then son,' appealed his mother, feigning reassurance. 'Let's not jump to conclusions. The army has its processes. After nearly five years of war, surely by now they have an established line of communication in order to report anything untoward that might happen to the soldiers.'

'That's for damn sure, 'expostulated Alan. 'You'd think that with the 60 000 Australians who've lost their lives in this imperialistic massacre, that the powers that be have refined their processes of informing next of kin about the slaughter of their loved ones. They've had plenty of practice! If that's the case, how come Arthur is reported dead when no one at the barracks knows anything about it?'

'Because it's a mistake son. Be patient and it'll all come out in the wash.' Hannah uttered with no great certainty.

Undaunted Alan continued. 'I even went to the *Sydney Morning Herald's* offices to see if they could tell me how they came by the information, but nobody's any the wiser there. They reckon that they get their casualty lists from Base Records, Melbourne so it looks like they might know the answer.' At that stage their conversation was interrupted by a tentative knock at the front door. Instinctively Hannah moved to respond. As she opened the door, her jaw dropped at the cluster of neighbours assembled before her. Mrs Bowles and her daughter, Joy smiled sympathetically, while Mr Greenshaw removed his hat and handed her a wispy posy of pansies.

'We were so sorry to read of your Arthur's death in this morning's Herald' he whispered. 'If there's anything....'

'Thank you all for your very kind thoughts,' interjected Hannah 'We think someone's made a mistake. Arthur is on his way to England on the HMAT SS *Boonah*. As you know he is in the army medical corps and he's accompanying a thousand of the latest reinforcements who were to go to the front. His ship the *Boonah* was the last to leave Australia before the Armistice was declared. He could be on his way home for all we know,' she offered with a blithe shrug. 'Alan's been to the barracks and they have had no news about Arthur or the ship, so I guess no news is good news!' A telegram has been sent to Melbourne to confirm that he is safe and sound. It was good of you to come, but don't worry!' she affirmed as she gently closed the door.

With no more information forthcoming that day, all that the Thwaites could do was wait. For the next two days Alan frequented the Victoria Barracks at Paddington and contacted his local MP in the hope of expediting some good news. No clergyman or ominous telegram arrived at their house, lulling both mother and son into a quiet optimism.

What neither knew was that a flurry of telegrams was crossing the continent enquiring about the well being of Sergeant Thwaites. Intelligence had informed Base Records, Melbourne, that the HMAT SS *Boonah* had been rerouted back to Fremantle from Capetown once news of the Armistice had reached the ship. At Capetown, she had taken on coal and begun her return journey across the blazing heat of the Indian Ocean. Base Records Melbourne, upon receipt of the cable from the Sydney Barracks, urgently dispatched a telegram to the Commandant of the 5th Military District in Western Australia. An enquiry was made as to whether they had received any news about Sergeant Thwaites. It was confirmed that the embarkation roll had listed him as the ship's dispenser when they left Adelaide on 22 October. The brief reply informed Base Records, Melbourne that the SS *Boonah* had returned to Western Australia but was currently in quarantine at Woodman's Point, just off Freemantle. Concerns were expressed about an outbreak of a contagious infection which may have been transmitted by the coaling party in Capetown. Investigation into the matter would be made and details forwarded when the quarantine order had been lifted.

More probing enquiries from Base Records, Melbourne, established that the quarantine order had been imposed on 11 December when the ship was just off Woodman's Point, Western Australia. The Officer In Charge of Troops, Major O'Halloran had developed a severe dose of the Spanish Influenza which was ravaging Europe. For fear of the disease spreading to Australia, it was decided that the patient had to be nursed in isolation and the vessel fumigated before anyone could disembark. Initially placated by those details and the likelihood that Sergeant Thwaites was confined by those orders, the administration of both military establishments at Paddington and Melbourne sought to allay the fears of the soldier's mother and brother.

However their relief was short lived. On 20 December an official letter stamped *On His Majesty's Service* was delivered to Mrs Hannah Thwaites. In a rush of expectation that such a missive could only herald good news, it was unfolded to disclose bewildering details and register unimagined shock. The letter dated 12 December was written from the HMAT SS *Boonah* in waters off Freemantle. It read:

> Dear Mrs Thwaites
>
> By this time you have probably received official notification of the death of your son, Sergeant A.C.Thwaites, but it is felt that you should have some expression of the general sense of regret and sorrow which this sad event has created among the officers and men of the reinforcements, recently returned to Australia on the above named troopship.
>
> Those who came into frequent contact with your son were truly impressed with his keen interest in his work on the AMC staff. His readiness and ability in his care of the sick made him as valuable as

his manner made him acceptable to those to whom he ministered. All ranks were distressed that the unfortunate malady which broke out on our homeward journey took the fatal form it did in his case, but the fevered brain was evidently outweighted, and the cord of life just snapped under the strain.

Will you and your sorrowing family and friends please accept this expression of the esteem in which your boy was held amongst his fellow soldiers and our earnest prayer that you may be divinely sustained in your great loss.

Yours sincerely

Captain L Griffiths

Officer Commanding Troops.

Stunned beyond words, Mrs Thwaites fell into the nearest armchair. That was where Alan found his mother upon his return from work. Speechless, but with a face drowned in tears, she held out the dreadful missive to her remaining son. Quickly scanning it, his worst fears were verified, but the elegant prose posed more questions than answers. Crunching the letter in an angry fist and pacing the room, Alan railed. 'So where's the official notification mum? How come we get this drivel before any official explanation? How did Arthur die? We know nothing!' He fumed. 'How dare they allow a condolence letter to hit us before Arthur's death is established and confirmed!' Frustration and fury temporarily overcame Alan's grief. 'The bastards won't fob me off this time mum,' he thundered. 'I'm off to the barracks, and I'll shove this letter in their bloody faces until I discover the true story behind this cock up.'

He received no protest from the slumped form of his sobbing mother. Holding her tightly, he kissed her lovingly and knew that he could not leave her in this state. Quelling the bile rising in his stomach he decided to temper the panic and moved to the kitchen to make them both a cup of tea, the universal balm for every crisis.

The next morning, Alan's entreaties to their local MP, Senator Gardiner galvanised an immediate response from the military authorities in Melbourne and then Western Australia. Questions were asked as to why the regular notification of death in terms of AIF order number 32 were not followed upon the death of Sergeant Thwaites. Enquiries revealed that the task was the duty of Major O'Halloran, but he had taken ill before despatching the formal details. It was only by chance that his subordinate, Captain Griffiths realised that no documentation had been forwarded, whereupon he undertook the task before writing personally to Mrs Thwaites. This explanation was presented to the family just before Christmas. It was not the Yule tidings for which they had hoped.

The question still remained as to how Sergeant Thwaites had met his end. Was he another victim of the virulent Spanish Influenza? An enquiry at Senate level revealed documents pertaining to an Official Enquiry conducted aboard the SS *Boonah* at Sea on 10 December. It was a sad and tragic tale for which no blame could be apportioned to anyone. Apparently Sergeant Thwaites had taken ill with influenza on the return from Capetown. Fearing the spread of disease, he had been hospitalised and nursed aboard the ship. His temperature and delirium had raged for days, reaching at its worst at 104°F. It was reported that at about 2200 hours on the night of 9 December that a number of troops were sleeping on deck, as the heat below was insufferable. The sentry had seen a figure cloaked in a blanket or dressing gown silently moving towards the stern of the ship. At the time no one recognised the person or took much interest, but in the next instant the slumbering men identified with alarm an ominous splash. Immediately the call of 'man overboard' echoed around the deck, before the Captain began to turn the ship to port in search of the soldier. Life buoys were hurled into the inky waters and lights shone upon the foaming waves but no trace of the man was found. The general consensus was that the man had either died of shock as he plunged into the icy waters or had been dragged under by the stern propellers. After about a half an hour of scanning the ocean, it was determined that the man had not survived and the ship was forced to resume its course. A roll call quickly established that the man overboard was Sergeant Thwaites. The subsequent enquiry convened the next day concluded a verdict of 'Suicide as a consequence of delirium with *pyrexia* and that no one was to blame.' Although Sergeant Thwaites was a quiet and often withdrawn personality, there was no evidence of contemplated suicide.

Such a litany of unfortunate events clearly undermined the normal channels of communication. The sickness of Sergeant Thwaites, his delirious suicide, the diagnosis of Spanish Influenza in Major O'Halloran as OC in charge of troops which rendered him too ill to complete appropriate documents for Base Records, Captain Griffith's kindly and compassionate letter to Mrs Thwaites, all triggered a measure of chaos before the details of death were finalised. By April 1919, the sad litany of events finally clarified and confirmed for the Thwaites the tragic circumstances of their son and brother's death. No one though was able to establish how the *Sydney Morning Herald* came by the information that it published about Sergeant Thwaite's death on 11 December 1918.

JUST REMINISCING

The events of this story are true, but names have been altered. The story reflects the larrikin humour of the Australian troops under adversity.

'Jeez we had some laughs amid all the shit we had to endure,' exclaimed Barney as he shuffled the last shreds of tobacco into the thin rectangle of paper. Balancing both items deftly in his left hand he licked the edge of the paper and rolled his cigarette.

'Bloody hell, if we 'adn't laughed we'd 'ave pissed ourselves,' retorted Fred.

'Talkin' about shit, remember that pommy bastard, the young junior officer who looked like he'd not been long out of nappies,' added Simmo.

The three men grinned and drew a long draught as they inhaled the familiar aroma of beer and cigarette smoke that permeated their once a year watering hole near the shrine.

'How could ya' ever forget that little trumped up turd who was in charge of our training division at Perham Downs?' snorted Barney. 'He was a real arsehole. Thought he knew it all and he knew bugger all when put to the test.'

'2nd Lieutenant Haughten, none other,' mimicked Simmo in a plummy English accent. 'Couldn't organise a one way queue to a country dunny if his life depended on it.'

'Jesus bloody Christ, when I think of that spruced up bastard in his shiny Sam Brown and solitary pip bought with family money, I want to puke,' interjected Barney.

'Yeah,' snarled Fred. 'I hated the way he always called us on parade straight after a 20 mile route march and then had the nerve to crime us on the state of our kit and uniform. I could look bloody immaculate if I had sat in the mess all day reading up on fuckin' regulations.'

'Those pommy officers were all tarred with the same brush. Stuck up pricks.' added Simmo. 'They never figured us Aussies out did they? All that snapping to attention and saluting. Calling us insubordinate because we thought all that parade ground stuff was a load of bullshit. You'd think that after four years of war they might have figured that you had to earn respect, rather than demanding submission through discipline or intimidation. I don't reckon that there would have been a bloke in the AIF who came home with a clean sheet. Yet they never understood the initiative our blokes demonstrated in the bloody war.'

'Too true. What did you get crimed for Simmo?' asked Barney with a wry smile.

'Just about every minor infringement in the book mate. Dirty kit, gambling, failure to salute and of course the usual AWL. A twelve hour leave pass never gave you much time to get into any mischief and those lovely girls in the English pubs deserved much more of my time than that.' he chuckled. 'I copped the lot from that bastard Haughten. Same old story in France. One time over there I met up with this gorgeous sheila. She served in the estaminet, and for whatever reason she seemed to like my ugly dial. It didn't matter that she couldn't speak our lingo, nor me hers 'ços we both knew what we wanted. Needless to say, when the grog ran out we wandered back to her house. God was that a great night! Clean sheets, comfy bed and a voluptuous mademoiselle. In spite of the shells, I reckon that night was the closest to heaven I got in the whole damn war. Anyway, when I woke the next morning I realised that I'd overstayed my pass by six hours. Looking out the window, all I could see was rain and mud. I thought to meself, I'm on a good wicket here. 'Aven't 'ad a day in bed since I was a kid. I might as well get hung for a sheep as a lamb and so I hopped back into the cot with Lisette. Yea' that was her name! Anyway, when she had to get dressed that night to go back to work, I did likewise. Had no option but to make me way back to camp. Naturally there were consequences. Field punishment number two and the loss of 28 day's pay.

That kept me sober and straight for a while.'

'Yeah, those field punishments were tough,' agreed Fred. 'I suppose they had to be considering we were at the front. At least they were handed out by our officers rather than the likes of 2nd Lieutenant Haughten. You know I reckon that bastard delighted in criming us or making life bloody miserable. Hauling us out of bed at 0200 hours for a parade, endless marching around that bloody parade ground and then making us stand at attention for hours under the pretext that there was to be a review by some military big wig. What a load of crap. The training battalion to him was just a big power game, an ego trip.'

Barney and Simmo nodded in assent and drew a deep drag on their smokes.

'He was always getting stuck into me for not having my pockets and pouches securely fastened,' smirked Barney.

'One day he was so pissed off with me that he sent me back to the hut and made me strip off.'

'What the hell did he do that for?' enquired Simmo. 'I didn't think the bastard was a queer.'

'I was a bit worried there for a minute mate,' retorted Barney. 'But no, he wanted to put me through a little exercise to remind me how many pockets and pouches I had to fasten properly. So there I was kitting up from undies to full combat uniform, pack, ammunition, medical kit, rations rifle and entrenching tool. He treated me like a kindergarten kid. Did I feel a dickhead!

'I remember that,' roared Fred, slapping his hand on his thigh. 'But boy did you have the last laugh in the end.'

'Sure did,' grinned Barney. 'It was a pity that I had to keep the payback a secret, because if I shared the details among the boys, sure as eggs some bastard would have leaked it.

'Whadya do mate?' queried Simmo.

'It's a bit of a long story but.....'

'Tell 'im Barney,' urged Fred.

Needing no more encouragement, Barney began to unfold his closely held tale of vengeance. 'Well,' he sniffed before taking a long sip of beer. 'It occurred on the night of our last pack march at Perham Downs, a week before we were to be deployed to France. Do youse remember? It was a filthy night, just on freezing, with rain pissing down when the illustrious 2nd Lieutenant Haughten decided to toughen us up for battle conditions. There we were, paraded in full kit at 2300 hours. Each platoon had to reach a specific check point by 0600 hours. Haughten gave his usual spiel, especially the need to look after all of our kit, to secure pockets and pouches so that nothing was lost. Just before we were sent off, Colonel Hannaford appeared. He had got wind of the fact that we loathed Haughten and I suspect that he didn't think much of him either. Anyway, he ordered Haughten to kit up and of all things accompany my platoon. Neither the men, nor our sergeant were happy, but from the look on his face neither was Haughton.

Well, there was nothing we could do about it, but begin trudging through the darkness, navigating with map and compass through blinding rain. Nobody spoke unless they had to and there was none of the usual joking and chiacking that goes on among the men. Everyone was pissed off over the pointless exercise, but worse still we had to tolerate that bastard in our midst. We were all hoping he would break a leg, but instead he plodded on, giving advice and directions non stop. He had that irritating upper crust pommy accent and that jolly jolly tone, that implied we were all having a wonderful adventure. What a load of bull! All we wanted to do was shove his bloody map in his fuckin' mouth and slit his throat! After slogging it out, by 0200 hours we had climbed near the top of a rock strewn mountain and even Haughten agreed that we needed to take a break. So, huddled under an overhang, we broke for a rest and a bite as we reassessed our orientation.

It was at this stage that Haughten politely excused himself and headed off among the rocks. One of the blokes asked what was he doing, to which I replied, 'He's going for a shit! With all this exercise his gut's probably in overdrive!' The boys laughed and continued eating and chatting, happy to be out of Haughten's earshot. Dunno why, but I decided to follow the bastard. It was easy to creep up on him, as the noise from the rain was almost deafening. Sure enough I soon found him, squatting gingerly amid the downpour with his strides around his ankles and his pack and kit wobbling on his back. 'Well old son,' I thought to myself. 'Here's your chance.'

'Now I digress before I go on. You blokes know, how, when you have a shit in the field, you instinctively look at it when you've finished your business and before you pull up your strides.'

'Yeah!' the other two men affirmed with amusement and anticipation.

'Well, knowing that,' continued Barney with relish, 'I just thought I would create a bit of mayhem for our illustrious 2nd Lieutenant Haughten. So while he was squatting there backing one out, I gently removed my entrenching tool from its strap and placed the blade discreetly behind Haughten's arse. It proved a fine receptacle and when he had finished his business, I quickly and quietly withdrew the spade.'

At this stage Fred and Simmo were grinning and goggle eyed, wondering what on earth happened next. 'Go on,' they urged, taking a large swill of beer and wiping the froth from their lips with the back of their hands.

'Anyway, true to the script, when he was finished, Haughten took a butcher's hook behind him before pulling up his strides. Needless to say, he couldn't see a thing. The look on his face was one of pure horror. So there, amid the pelting rain and freezing wind he started to panic. He must have thought, "Where the hell is that bloody turd?"

So off comes the pack and then he begins to fossick around his strides. Of course there was nothing to find.'

At this stage, Fred and Simmo were laughing so much that they almost choked on their beer.

'In a frenzy he must have thought that he perhaps needed to shake out his strides, but before he could take them off, he had to remove his boots. So next thing you see, is the bastard standing in the raw from the waist down shaking his strides in the rain looking for the invisible turd. I was totally cracked up and thought I would burst with laughter. As if that wasn't funny enough one of the buttons on his pocket must have burst with all the shaking and the next thing Haughten sees, is his pens, compass and whistle flung among the mud. What was that about secure fastenings for pockets and pouches I remember thinking?'

'Christ Barney, how did you hold it together ?' roared Fred. 'What happened next?'

'Well with military dignity crumbling around him and almost manic with frustration, Haughten had no option but to put on the strides, still uncertain as to what lay inside them. Then of course he laboriously squeezed into the soggy boots before scrambling around the ground to retrieve his belongings. By this stage he was soaked and looked a totally dishevelled mess. While he was scrambling around in the mud I decided to take my leave. When I returned to the boys I innocently asked where on earth Haughten could be. Needless to say there were all sorts of suggestions as to where they hoped he'd be, but no one had a clue as to why he'd been gone so long. After another ten minutes trying to make the most of the shelter in the torrential rain, a bedraggled Haughten appeared. It was a memorable sight lads. A real sight for sore eyes. The once impeccable, 2nd Lieutenant Haughten looked as goddam awful as the rest of us. Furthermore, he was walking very gingerly, still uncertain about what part of his clothing might be housing the lost turd. He had given us the shits for that whole month's training and it was sweet relief to give him some of his own back. Better still, was his silence for the remainder of the exercise. Only I knew what he was probably thinking .' he chuckled.

'Crikey Barney, you deserve a bloody medal for that bit of bastardry.' exhorted Simmo in amazement.

'Sadly,' ruminated Barney with sigh, 'Initiative doesn't earn you a gong mate. If it did, the Aussie ships would have sunk on the voyage home with all the metal they would have collected.'

A COUNTRY CENOTAPH

A loss of any family member in conflict results in devastating grief. This story explores the impact on a family ravaged by multiple losses. It highlights that small country memorials are worth more than a fleeting glance as they record the devastation war so often wrought on a small community. The names and events of the story are all true.

When John Williams married Margaret Middleton in Dunkeld, Victoria in 1886, their respective families had farmed in the region for nearly 20 years. Their parents had been drawn by the lush pasture around the little hamlet that lay at the foot of the southern edges of the Grampians. The Grampians were a single string of low lying ranges that broke the horizon of the rich plains of what became known as the Western District. Like their forebears, the young couple hoped to run sheep and establish a fine legacy for their children to inherit.

As devout Roman Catholics they were sure they would be blessed with a sizable brood and indeed they were. With five strapping sons in eleven years, John felt a certain future. Yet that was not all. Over the next fifteen years, by some miraculous magic, the scales were balanced with the arrival of four hale and healthy daughters. Lord be praised they often exclaimed, and indeed he was. The Williams were a happy family bound to the church and their small community. The boys worked alongside their father while the young girls looked after each other and stayed close to the timber cottage. However, pleasures were few and with eleven mouths to feed, the days were long and the work endless for both parents. Although they paid their way, due to the strong market for wool, both knew that their real wealth rested in their children and in particular the five sons who would inherit the land and the future it offered.

As they grew from adolescence into young adulthood, one by one, the boys began to grow restless. They longed for a life beyond the small valley and gradually each struck out to define themselves beyond the family circle. Robert, the eldest was the first to leave home to head for Port Augusta in South Australia. As a seaport and railway junction, work was plentiful for a strong young man. His example was followed by Frank who also found ready work there as a labourer. Young John, his father's namesake, headed north, lured by the exotica of the tropics, only to find that Queensland needed labourers too. Michael chose a different pathway. Since his first glimpse of the sea as a small child, he knew that he could not resist its allure. As a handyman around the farm he loved to tinker with machinery. In response to both his passion and his natural inclination, he farewelled the farm to join the newly emerging Royal Australian Navy. The youngest of the five brothers, Peter, found his way to Queenstown in Tasmania, where he quickly secured work in the mines.

John understood his sons' exodus, but was devastated that rampant hormones and the allure of attractive women overwhelmed common sense. Margaret in turn was bitterly disappointed that none of their five sons wanted to embrace the opportunity on their own doorstep. She missed their steady banter and their young strong, bodies around the farm. For all of the family, life suddenly became much harder. By 1914 John and Margaret had four daughters under ten to care for but they still hoped that the boys, having sown their wild oats, would return to the fold. Tragically the outbreak of the First World War and Australia's decision to support the Empire dashed such silent longing. Patriotism spread across the country like water from a ruptured dam. As soon as the call for young men to enlist became public knowledge, both feared that their lads would succumb to the excitement of travel, adventure and six bob a day. They were not wrong.

At age 17 Michael had enlisted in the Royal Australian Navy in 1912 for a term of seven years. He therefore had an automatic entry into the conflict. The newly developed submarines had piqued his passion and he was quick to seek a transfer to the RAN submarine fleet which consisted of two of these new vessels. He became a stoker on the *AE2* which was commissioned in 1914. In a deployment that remains largely unknown to Australians, Michael was aboard the *AE2* as a part of the Australian and Military Expedition Force that was quickly despatched to Papua and New Guinea to quell German imperialist ambitions in the region. After that mission, the *AE2* was directed to Europe. It was towed for a large part of the journey to preserve its condition. There, based in the Mediterranean, it became one of the fleet of vessels endeavouring to support the Gallipoli landings. It was in fact the only allied vessel to breach the Dardenelles and silently enter the sea of Marmara, where it followed orders to 'run amok' and deliver as much damage as possible. Unfortunately, on 30 April 1915, mechanical problems forced the submarine to surface, whereupon the entire crew was captured by the *Sultanhiser* and the submarine scuttled upon the captain's orders.

Stoker Michael Wright Williams was one of the 35 member crew. He was imprisoned in a number of Turkish prisoner of war camps and proved himself to be anything but a compliant POW. He refused to work and being identified as a trouble maker was removed from camps at Pozanti and Bahce where conditions deteriorated and he was subsequently isolated from his mates. In Bahce a rock slide smashed the men's sleeping quarters and Williams was badly injured. After that, what happened to him remains a mystery. It is speculated that he died of dysentery or malaria on or about 29 September 1915. He was one of the four crew members of the *AE2* who died in captivity.

In the interim, his 23 year old brother John Edward, had enlisted on 18 September 1914 at Bundaberg in Queensland. As a member of the 14th Battalion, he sailed from Melbourne on the HMAT *Ceramic* on 22 December 1914. Unable to leave the farm and family, not to mention provide for the expense of such a journey, his parents were not to know that they had farewelled their son years before. Like thousands of other members of the First AIF, John suffered the rigours of training in Egypt. There, he was hospitalised at Heliopolis with a septic hand before being deployed to the Gallipoli Peninsula.

A bullet wound to his foot demanded a brief return to Egypt and the comfort of hospitalisation at Alexandria. Such respite and recuperation were short lived and John rejoined his unit on 20 July 1915. His next battle was the major assault on Lone Pine, where he was killed in action on 8 August 1915.

Shattered by the news of Michael's capture and John's death, Margaret asked the local school teacher, Charles Overman to make enquiries on the family's behalf. He wrote first to Senator Pearce as Minister for Defence seeking clarification and further details of the Lone Pine disaster. Sadly no news was forthcoming. With broken hearts and ageing limbs both parents realised that they had to forsake their beloved farm. It was hoped that the small township of Horsham would provide security and access to the medical treatment that both were starting to need.

Their sense of dislocation and anxiety were exacerbated when their son Frank William, wrote to them of his enlistment. He had enlisted as a member of the 10th Battalion in Adelaide on 2 August 1915. This was the first of the South Australian Battalions to be formed. After Training at Mitcham and before departing for the Middle East on 27 February 1916, he travelled to Horsham to bid his parents a loving farewell. Within six months of basic training in South Australia, Frank embarked on HMAT *Benalla* bound for the Middle East. In the hot desert sands, he was toughened up before being deployed via Marseilles to the Western Front. Here, he too was killed in action on 23 July 1916, just days after the horrendous battle at Fromelles. He was just 23 years old.

Unbeknown to his parents, John and Margaret's youngest son, Thomas Peter, also felt compelled to join the tsunami of volunteers. Enlisting in Claremont, Tasmania on 10 November 1916, he was directed to Broadmeadows on the outskirts of Melbourne for his basic training. His parents knew nothing of his enlistment or embarkation on the *A34 Persic* from Melbourne on 22 December 1916. Military life was not easy for Thomas and on the voyage to Cape Town he was disciplined for insubordination and gambling. At Cape Town he was offloaded and returned to Australia suffering from Venereal Disease. Given the stigma of the disease, Thomas never communicated with his parents on this ignominious return to Australia. Following treatment, his next point of departure was from Sydney, aboard HMAT *Port Melbourne* bound for Liverpool and the training camps in England. Here his service record was singularly unimpressive, with punishment for a litany of minor offences and periods of hospitalisation. Finally, on 13 March 1918, Thomas was posted to the 14th Battalion in France. As a member of the great spring offensive against the German military might, and so close to the end of the Great War, Thomas' combat ended with a gunshot wound to the head on April 18. Later that day he died of wounds.

Tragically, four out of five of the Williams boys had been claimed by the maw of the Great War. The pain of such loss is unimaginable and by and large John and Margaret kept their grief to themselves. Obviously both parents longed to know where their boys were buried and made enquiries as to the possible return of any belongings as keepsakes. Only Thomas had a known grave at the Doullens Communal Cemetery in France. Michael, John and Frank rest in unknown graves in Turkey, Gallipoli and France. In a cruel twist of fate Thomas's meagre effects did not ever reach Australian shores, owing to the sinking of the transport ship, SS *Barunga*.

In the immediate years after the war the Williams family received their sons' campaign medals and memorial plaques honouring their service. Without their sons' allotments though, life was even crueller. In both desperation and despair Margaret Williams took it upon herself to entreat support and compensation from Prime Minister Billy Hughes. Her letter was dated 1919 and she wrote:

> Sir
>
> I put my case before you. I have lost my four sons at the front. Pte. John, killed in action at Lone Pine 1915, Pte. Frank killed in action at Somme, 1916, Stoker Michael, Late AE2 submarine died POW at Pozanti, 1916 and Pte. Thomas died of wounds France 1918.
>
> Sir, It's Stoker Michael I want to put you about. I got paid up to June

1916. The Turks never reported his death. I can't get his pay from June as I have cards he wrote me in September 1916.

I received his cards regular until 10th September was the last one he wrote to me. I have one son married left and four girls, the eldest being 15 years down to 7 years.

We left Dunkeld as I could not live in the bush under the terrible loss of my dear boys and we were 20 miles from the Dr. and my husband wants medical attention and also one of my girls is not strong.

Sir, I think you will find my case a bad one as brothers and nephews have all done their duty.

Sir, I just heard that my son who died a POW met with foul play. They blame the warders in the hospital and my son was buried somewhere in Turkey which is hard on free Britishers.

I will close hoping you will look into my case.

I remain yours truly

M.M.Williams.

Mrs Williams never received a reply from the Prime Minister and so the family was left with grief and penury. Sadly, in a later move to the city and the death of John Williams in 1928, he never lived to see the elegant and solemn memorial erected in honour of the young men of Dunkeld who never returned. The residents of tiny Dunkeld contributed the princely sum £625 for the memorial which was unveiled on 4 August 1929. It depicts a sombre statue of a digger with Arms Reversed who stands atop a plinth bearing the names of the fallen. In gold lettering, carved into stone, among others, the memorial pays tribute to the four Williams boys and the huge sacrifice of one Australian family. At the end of the inscription the memorial says: *We lie dead.* It alludes to not only the soldiers, but the longing and hopes of their parents who would never see their seed, harvest the lush pastures of their dreaming.

IF ONLY SHE KNEW

This story is a composite of the recollections of my great aunt and the file of Sergeant John Birch. It details one of the many wartime marriages that evolved between nurses and patients. It also explores the fact that not even one's nearest and dearest could ever fully comprehend what the men had endured and achieved.

Jack Birch was a quiet, self effacing man not given to fuss, panic or excess emotion. When pruning one day in his garden, he found that the air in his lungs could not match his effort and feeling dizzy he sat down under the apricot tree and silently died. He had been the perfect foil for my great aunt Mattie, whose red hair and quick Scottish temper constantly animated her tiny frame. She was furious that Jack had seemingly chosen to die, leaving her with a big weatherboard house to maintain and a only half a pension on which to do it.

For as long as my mother could remember, her Auntie Mattie had treated her husband with a degree of bitter disdain. For years he had been banished from the marital bed. Instead he slept on an old iron bed under the house, in a space that had been cut into the slope of the land. As a child I was told that Uncle Jack slept there to escape the summer heat, but it was not always hot in Melbourne. As he was often wracked with violent coughing fits, I often wondered how he coped with the chill of winter. It was not something that Auntie Mattie appeared too concerned about, so I grew to understand that the arrangement must have been satisfactory to both.

Even though Jack was out of sight at night, his daily presence was enough to evoke Auntie's vitriol. In retirement, he pottered around the garden and ate his meals with a gentle 'thank you' which seemed the only discourse he shared with his wife. From Mattie's perspective, her husband had no conversation and seemed to exist in a kind of twilight zone of his daily reverie. Jack initiated no activity such as a film outing or a trip to the city. According to Mattie, he lacked ambition and was suffused with the laconic ennui that afflicted all Australians. The only occasion that would draw him away from his plot of land and to exercise his vocal cords was a visit from Mattie's sister Belle and her husband Fred. In such company, Jack would engage in men's conversation about the football or racing, while Auntie Mattie would regale my grandmother with laments of the blighted expectations that Jack Birch had failed to deliver.

Auntie Mattie had been born in Edinburgh and was one of ten children to William and Isabella Lauriston. Her father was a wheelwright and her mother had died not long after the tenth birth. Forced to earn her living at an early age and contribute to the household economy, Mattie, or Martha as she was formally known, had followed in her older sisters' footsteps and become an accomplished seamstress. Yet with the outbreak of the First World War and the dreadful casualty lists that followed, she felt drawn to serve, just like her three brothers who had readily enlisted. Casting aside her fashionable art, Mattie decided to undertake work as a Volunteer Aid Detachment (VAD) Nurse. Like thousands of other young women at the time, she quickly learned the rudiments of first aid and bandaging, but life as a VAD was by no means a glamorous role. The work was gruelling and physical, with long shifts regimented by military discipline. Sweeping, cleaning and polishing under the watchful supervision of a Registered Nurse, saw Mattie reduced to the ranks of little more than a glorified skivvy. Nimble, delicate hands accustomed to fingering silk and sequins were transformed into calloused claws that scrubbed bowls of vomit, blood and shit. Yet an old photograph of Mattie as a gentle, fresh faced young woman in her VAD uniform, with the red cross emblazoned on her chest, reflects a surprising aura of calm and conviction of purpose. She actually looked happy.

By 1916 Mattie had completed her basic training in her home town of Glasgow and was soon transferred to the 2/1st Stationary General Hospital in Birmingham. Scottish and Australian troops seemed to make up the bulk of the patients, and the carnage that she confronted daily soon caused her to realise that her work was not for the faint hearted. Initially, she was confronted with having to sponge men's naked and broken bodies and dispose of their foul wastes. Inevitably, the stench of suppurating wounds, pus, blood and faeces caused her own bile to rise into the foul bowls that she was constantly forced to empty. Yet the shock of her brother, Hamilton's death in the calamitous annihilation on the Somme and the brief letters that she had received from her brothers William and Frederick, provided a constant reminder of what these poor men had endured on the battlefield. She could not help but imagine her surviving brothers wounded and hospitalised, and with them always at the front of her mind, she resolutely steeled herself to every bitter task.

As the months ticked by and the faces in the beds constantly changed, Mattie's confidence and responsibilities grew. No longer afraid and embarrassed about the flesh she nursed, she began to realise that the best she could offer these men was more than physical care. She marvelled at the courage of the bonny wee Scots lads and the large limbed Australians who spoke with a seductive, sleepy drawl. The Scottish boys loved the familiarity of her brogue, whilst the Australians kept her talking in order to savour the lilt of her gentle dialect. Clearly all of the men were hungry for conversation. In spite of their wounds, laughter rippled among them and in time she felt herself relishing the witty banter that resounded through the ward. The jokes and wry asides reminded her of her brothers and the joy that they had shared as a family. The senior sister constantly warned Mattie about becoming too familiar with her patients, but when publicly rebuked, the men's strident objections soon saw sister turn a blind eye.

In her daily chatter with the men, Mattie was fascinated to learn what each had done in their lives before the war. There were of course a fair sprinkling of labourers, but it was the writers, artists, printers and teachers who had a profound impact upon her. Thinking of them collectively she began to grasp the enormity of the war's legacy. In her care were hundreds of gifted men whose talents faced oblivion, whose seed would never lay the foundations upon which the future would be shaped and crafted. In this reflective vein she thought of her own brothers and wondered whether the surviving two would have their chance to carry the Lauriston name into the next century. Bitterly, she grieved Hamilton's death which had achieved little more than to fertilise the fields of France with blood and bone. All men had a right to leave their mark she determined, and thus resolved, Mattie began to activate what was once a fleeting thought.

The idea was deceptively simple and even Sister Scott saw no problem with it. As she grew more comfortable with individual patients, Mattie produced a beautiful gilt leaf autograph book with polished pristine pages. In her soft, beguiling manner she worked her way around the ward asking every man who was able, if he would like to write, paint or draw something that would live on, long after the men had been discharged or returned to the trenches. She was surprised at the alacrity with which the project was accepted and absolutely amazed at the time and patience the men invested in the task. One man asked for some watercolours and with a minute brush painted an evocative scene of Ypres after the initial bombing. He was happy to sign it, but would not write what he said after having handed the book back. 'The bloody Poms,' he lamented. 'Great battle plans as long as they could send the Scots in first as shock troops. The bastards!'

J.S.White of the Mcleod Regiment offered a more whimsical approach. He plucked out a wisp of his hair which he stuck to the page with a piece of sticking plaster. Underneath it he wrote,

'You ask me for something original
Something out of my head.
As I haven't got anything in it,
I'll give you what's on it instead.'

Gradually the gilt pages of the tiny book were filled with mementos and *bon mots* from a variety of the convalescents in Mattie's ward. It became a possession that she kept until her dying day as a memento of so many of those brave lads for whom tomorrow never came. Understandably amid the horror and sadness, Mattie wondered about her own future and almost as an epilogue, she added her own whimsical thoughts on the last page of the book.

> 'Somewhere there waiteth in this world of ours
> For each lonely soul, another lonely soul.
> Each chasing each, thro' all the lonely hours
> And meeting strangely at some sudden goal.'

Little did Mattie imagine that her strange meeting would take place on the very ward in which she toiled so arduously. By 1918 the German spring offensive was in full swing and the casualties continued to mount as the allies fought back resolutely. The Australians had proven a formidable opponent to the Germans and it was hundreds of their seriously wounded troops who were transported from Casualty Clearing Stations in France, to the Stationary General Hospital at Birmingham. By the time the wounded arrived at the hospital on 16 April 1918 their condition was appalling. Still covered in mud and blood with coarse field dressings on their wounds, the young VADs were directed to cut the uniforms off the men and wash them before the sisters and doctors would examine them. Amid the chaos of pain and putrefaction Mattie noticed a tall Australian lying quietly beneath a light that highlighted a grey face. She observed that he had three wound stripes on his muddied sleeve, and sympathetic to his previous suffering, resolved to strip his clothes immediately in order to expedite his treatment.

Cutting up the right leg of his bloodied khaki trousers Mattie saw the cavity in his hip which was packed with a large wad of putrid field dressing. Valiantly she endeavoured to make conversation as a kind of distraction from the soldier's pain but the wheezing that suggested a reply, indicated that a gas attack had also accompanied the bomb blast. Nonetheless, she persevered, removing his clothes and gently sponging weeks of accumulated filth and gore from his gaunt frame. Speech was beyond her patient, but she gleaned that he was a sergeant by the name of Birch and that his steely blue eyes expressed an abiding gratitude for her ministrations. She had never felt that silent appreciation before and his gaze touched the very core of her soul.

From that day on Mattie hurried to each shift, eager to see how the quiet Australian was progressing. Each day she marvelled at his resilience and each day she learned a little more about him. His Christian name was Joseph although he was known to all as Jack. He had been born in Greenwich 32 years ago and his family had migrated to Queensland in Northern Australia when he was a boy. He had thought that by enlisting in the AIF that he would have an opportunity to see the 'old country' of which his parents spoke fondly. However, from what he had seen of training camps at Perham Downs and Codford, and hospitals in London and Cardiff, he had cause to wonder why people venerated this bastion of the Empire. From his perspective, the motherland was blighted by the weather and doomed by an antiquated class system.

He said as much to Mattie as she made his bed every morning. Knowing nothing else, her initial response was to scoff at such criticism, but when he began to describe an alternative life she began to listen with greater interest. Jack described his property at Stanthorpe, Queensland, where the big open sky, shone like smiles at a children's picnic. Every day was warm and while it rained in the wet season, winter clothes were unknown. She was fascinated at his description of the family home which was built of timber on high stilts, designed to capture a cooling air flow. The fields, or as Jack called them, paddocks, fed grazing cattle and a few sheep. Life was governed by the cycles of the seasons and the animals rather than the clock. To Mattie, raised in the smog and crowded tenements of Glasgow and Edinburgh, the picture painted by Jack was as alluring and indelible as the entries in her autograph book.

Over the next three months of Jack's recovery and convalescence, Mattie grew more fond of the gentle Australian and envied the world he yearned for and loved. Unwittingly, both he and his descriptions of a life that marched to the beat of a different drum gradually seduced her. After his discharge in June, Jack wrote regularly from Longbridge Deverill where he was stationed. Marriage was mentioned, but distance and duty deferred her response. Yet in a melange of horror and delight, a letter that Mattie clasped to her bosom on 20 July informed her that Jack was to be posted again to France in early August. A decision had to be made and Jack had made it for her. He had organised the Vicar at Holy Trinity Anglican Church in Coventry to marry them on 28 July. At 32 years of age and facing a desolate post war future devoid of eligible bachelors, Jack's offer also came with the promise of a life in Australia. With that letter, Mattie knew that Jack Birch was the other lonely soul whom she had met strangely at that sudden goal.

It was only after Jack had returned to France, that Mattie discovered that he had gone AWOL for the week of their wedding and had consequently been reduced to the ranks. He never complained about that, nor the additional months of hospitalisation back in England after yet another serious and debilitating gas attack. He never shared his experiences about the battle that invalided him from the front and the blighty that fortuitously brought him back to her. Jack knew that his war was over and turned his thoughts to Mattie and a return to sunny Queensland. That dream sustained him until the Armistice and through his rehabilitation. Given the deaths of her remaining two brothers, Mattie's family circle was not only broken but irrevocably shattered. Europe was in ruins and Britain was sinking into a sea of poverty as quickly as the *Titanic*. With the promise of a bright future on the other side of the world, Jack and Martha Birch boarded the *Konigfred* on 20 June 1919, bound for the antipodes where castles and palaces seemingly populated the landscape.

Disillusionment was quick to consume Mattie. She found nothing regal about Queensland. The culture shock of her transition was something that she never adjusted to. The untamed country and the rough frontier towns, devoid of the trappings of civilisation as she knew it, left her feeling as if she had been marooned on Mars. Stanthorpe was little more than a collection of dilapidated wooden humpies, while the family home was akin to a shed on sticks. She loathed the humidity that coated her beautifully made dresses with mildew, the isolation and the horrible spiders and snakes that lived within and without the house. Her promised paradise was so totally removed from all that she had imagined, that her wrath bit like a viper at Jack's happiness. Yet, Jack never fought back. He deflected her harangues like a heat shield and quietly went about his daily tasks, while Mattie steamed and fumed herself almost into heat stroke. She could take little comfort in the son that she had delivered, as like his father he was introspective and inclined to disappear into the bush as if it was a friend that he must see.

In spite of Mattie's fury, Jack remained content. He never spoke to his wife of his exploits at the front, but instead was grateful to have survived and to have a son to carry the family name. Whatever made Mattie less ferocious and his life bearable, became a sought after reality that soon found its answer in a relocation to Melbourne. There Mattie had her sister Belle and family. Melbourne, which was once deemed 'marvellous' was a growing metropolis with elegant Victorian buildings, shops and public transport. Thankfully, the sale from the Stanthorpe property bought a neat house in a suburban street. It was devoid of snakes and black huntsman spiders.

Mattie had badgered Jack into this civilised relocation, yet after this move he withdrew more into himself. Silently, he longed for the smell of the eucalypts and the clatter of the wild birds. He missed his Queensland family too, but rationalised that that severance was just another compromise that a man had to make to keep the peace. Responsibility saw him assume a regular position with the State Electricity Commission and in spite of the nine to five routine, Jack endured it for the greater good. From his perspective, Mattie was never happy, but he accepted her rages about his failure to fulfil her exalted sense of entitlement as the price he had to pay for surviving the war and having a son. Nothing would appease her wrath. From her perspective, Jack had promised much and delivered little. He lacked ambition, gumption in fact. He was too passive about life. He needed to be bolder and take life by the horns! Although Jack listened to this cyclical tirade which echoed in his consciousness, he knew in his heart that he had been a devoted son, a dutiful soldier and a responsible family man. What more could he do in the service of both country and kin?

As he laboured in the garden that day, the branches of the apricot tree would not yield to the angle of his saw. Panting and sweating profusely, a coughing fit engulfed him and he flopped down in the long grass to catch his breath. Gradually the dizziness abated, replaced by a platoon of soldiers on parade. Facing it was Lieutenant Colonel Witham, about to pin a Military Medal to his breast. In doing so he began to read the citation, the words of which Jack had long since forgotten. The Colonel read in a stentorian voice that echoed in Jack's dwindling consciousness.

> 'This NCO, Joseph Birch, displayed magnificent conduct during operations in front of Verguier, North West of St Quentin on 15 September 1918. He was in charge of a platoon during an intense heavy artillery and gas bombardment and although blinded by gas, he continued to supervise the anti gas measures. By being led about by a private, he made sure that all reasonable precautions had been taken. It was not until he received a direct order that he was induced to go to a R.A.P. His bearing throughout the operation was of the highest order.'

The vision faded as darkness engulfed him. If only she knew.

LIBERTY HALL

During World War 11 Australia's defence was supported by an influx of American troops. They were immaculately groomed, possessed good manners and were generous with their wages. They presented a seductive charm to the less sophisticated Australians; especially the women. Their easy warmth, endeared them to many Australians who understood that their contribution in the Pacific was fundamental to Australia's survival

My mother always referred to her childhood home as Liberty Hall. I think that this was largely due to my grandparents' open hospitality. My grandfather was a genial man's man. At weekends he frequented the racetrack and the local pubs and it was not uncommon for him to invite his mates back home after closing time to continue the libations around the billiard table. My grandmother, as was in keeping with the times, never complained about Fred's many 'blowins', and always managed to rustle up some sandwiches and cake to soak up the alcohol and thus double the hospitality. On numerous celebratory occasions it was not uncommon for the bath to be filled with ice and brown long necks, to roll a Saturday afternoon into a Sunday morning. No wonder everyone loved to be invited back to Fred and Belle's.

Fred's favourite watering hole was McNamara's, a Caulfield pub that abutted the eponymous racecourse. It was here in 1943 that he began to chat with an immaculately groomed American Marine who was stationed nearby. Always impressed with the polite American troops supporting Australia in the Pacific war, Fred was quick to shout the fellow a beer, offer a welcome handshake and start up a conversation. Obviously, he quickly warmed to the young marine's ready smile, laconic mid western drawl as the two men began to chew the fat over the war and their respective families. The tall marine was 22 year old Corporal Jack Williams from

Canton in Illinois, who had enlisted in Peoria in 1940. He knew no-one in Australia, and figuring that he should experience some Aussie hospitality, Fred lost no time in inviting him back home after the pub had closed.

As my father was a member of the AIF stationed in Darwin, my mother, Joyce also shared the family domicile. No doubt both she and Belle were pleasantly surprised when Fred arrived home with the handsome and laconic marine. The American troops, in their meticulously tailored uniforms made the Australians look like scarecrows in the sacks of their AIF issue. This unfamiliar arrival at the back door not only shone like some emerging star, but sounded like a radio personality with his fine manners and engaging warmth. Belle and Joyce were instantly bedazzled. Naturally Jack's welcome immediately swung into full gear and a generous meal followed drinks during that first visit. Over the following months 'Lauriston,' was to become a regular haven for the homesick marine. Jack's laughter and gracious appreciation of my grandparents' hospitality immediately endeared him to my mother's siblings, John and Laurie. Unconsciously, for Fred and Belle, Jack became a surrogate son.

With each visit, there was no shortage of conversation around the dinner or billiard tables, after which a sing along around the piano invariably cruised the family into the darkening night. On occasions when my grandparents had finally retired to bed, my mother and Jack would perch beside the waning fire to discuss the horrors of the conflict that surrounded them, the loss, the uncertainty of the future, and perhaps the almost unspeakable possibility that either my father or Jack might lose their lives in the conflict. In moments of uncertainty one had to grasp the flickering wisps of human connection to nurture courage in order to scaffold hope for the future. Mum and Jack did just that.

Over many months, these long conversations established that Jack's family consisted of his father Gus, who had fought in WW1, mother Rachel and sister Maxine. Jack and his sister had attended the local Canton High School where he was a happy go-lucky member of the band. Upon leaving school he enlisted in the marines at Peoria, which of course as circumstances prescribed, led into an automatic enlistment in the Military Forces upon the United States entry into the war after the attack on Pearl Harbour. The war in the Pacific and Australia's precarious vulnerability to the Japanese, drew the Americans into an alliance which saw them deployed to Australia. Fred, Belle and indeed my mother, wrote to the Williams family, by way of introduction, and reassurance to them that their son was well and enjoying family life as best he could with the Coles.

Finally in December 1943 Jack received news that he and his battalion were to move out to an unknown posting 'somewhere' in the Pacific. His final night of leave was spent with my grandparents. A magnificent meal was provided by Belle, toasts to a safe return were offered and Auld Lang Syne sung, as Jack's surrogate Australian parents linked arms and hearts across generations and nations. With the final embraces at the gate, Jack snapped to attention and saluted his new found family whose tears streamed, as his tall figure strode purposefully to the station. With heavy hearts my grandparents waited at their front gate the next morning knowing that the troop train would pass their house on its way to Port Melbourne and embarkation. As the train chuffed up the hill before blowing its whistle at the little level crossing, eager faces and waving arms hung from the open windows. The marines whooped and whistled as the train puffed its laborious journey up the line. Hoping upon prayers and begging for one last sighting of their special marine, there was Jack in the last carriage waving and gesticulating as he threw a small package out into the trackside grass.

As the train disappeared amid smoke and distance, Fred crossed the road to retrieve the package. In it were two brief letters written in pencil on two pages torn out of a pocket notebook. The first was addressed to my mother and remained as one her 'treasures' for the rest of her life. The second was to Fred and Belle. Today these scraps remain torn, tattered and brown with age, but the sentiments of a young man who found a loving welcome so far from home transcend time and place. They echo the feelings of thousands of soldiers who found love and comfort amid the devastating dislocation of war.

> Dearest Joyce
>
> Sorry I was unable to get over to say so long. But perhaps it is better this way as you know how morbid people can get at times Joyce. I am really going to miss those little friendly discussions we had, even those little dashes for the train. I am really going to miss you a lot. Will try and write as soon as possible. Mine will take longer than yours to arrive. I should receive yours by the time we arrive if so send it at once. I do hope to see you soon. Tell Cam and all the rest of those swell people I send my best. If you can't read this you can blame the train this time. Please take care of yourself. Keep the old fire going in the fireplace and I will be back soon.
>
> Love Jack

The second note to my grandparents was equally cherished and it reflected the abiding bond that Jack had developed with my mother's family.

> Dearest Fred and Belle
>
> Well this looks as though one of the greatest chapters in my book has to close. Words can't express what is in my heart for you wonderful people. Fred, please take good care of Belle and the rest of the family-as you have a wife and family that would give any man just pride. And Belle I don't think Fred or you couldn't have made a better move 30 years ago.
>
> Please both of you write when you can and if I don't answer please don't think it because I don't want to. Fred tell the boys at the pub to have one for me every now and then and I know that I will get the best toast. Belle I know my mom would love to hear from you. Please take care of yourself. Once more I will try in an awkward way to say thanks.
>
> Love
>
> Jack

Needless to say my family had no idea that Jack and his battalion had been deployed to Papua and New Guinea. They never knew for many months that his engagement in the Battle of Cape Gloucester on the southern tip of the island of New Britain would cost him his life. The marines were a part of the operation codenamed 'Backhander,' a part of which was operation 'Cartwheel.' It formed the main part of the allied strategy in the South West Pacific area during 1943-1944. It was the second landing of the US Marine 1st Division following Guadalcanal. The objective of the operation was to capture the two Japanese airfields near Cape Gloucester that were defended by elements of the Japanese 17th Division.

The main attack began on 26 December 1943 when the US Marines landed on either side of the peninsula. A Japanese counter attack briefly slowed the advance but by the end of December the airfields had been captured and consolidated by the marines. Yet on 7 January 1944, Jack and his combat patrol had been ordered to clear the jungle area leading to the foothills. They were nearly two miles beyond the previously captured air strips, carefully stalking through dense jungle. It was here under the Volcanic Talawae and Tangi mountains that the retreating enemy had dug in.

Leading the marines into the dense jungle was the 'point' or scout, none other than Corporal Jack Williams. He whispered to the man next in line and the whisper travelled from man to man. 'Keep a five pace interval. Fresh tracks ahead.' Suddenly a machine gun burst sprayed the trail. The well camouflaged Japanese ambush had been exposed prematurely. As soon as Jack was hit, two other corporals rushed to his aid as he lay beside a fallen log. He was out of sight of the Japanese machine gun, but knowing his men were pushing forward, he raised himself up and waved the men back. The Japanese saw him raise up too. This time the Japanese machine gun opened up on him and ended his courageous stand with a hail of bullets. In sacrificing himself, Jack saved the lives of the rest of the patrol. As one newspaper reported; 'Jack had guts enough for a dozen men.'

News of Jack's death was not received by his family until 12 February 1944. His mother was prostrated from the shock, only having received a cheerful letter from her son the day before. Jack's sacrifice was hailed in the local newspaper;

JAPS KILL CANTON MARINE AS HE SAVES LIVES OF PALS.

But for the Williams family medals and mastheads proved no solace. I do not know how my mother and her family first heard the news but obviously they were quick to express their grief and deepest sympathy to Jack's parents. A letter dated 10 April 1944 remains another 'treasure' that my mother kept until the end of her days.

Dear Mrs Darke

Your letter arrived when I was in the hospital. On February 4th about 5 o'clock in the evening. I was all alone and a messenger boy came to the door with a telegram stating that Jack had been killed in action January 7th and you can imagine what that news did to me. A very sharp pain struck at the base of my brain running down my spine and I could not remember where Daddy and Maxine were. That was on Friday evening and I kept getting worse and the next Thursday eve they took me to Peoria Hospital and I was there five weeks. Oh! Mrs Darke, I will never get over his death as times goes on it gets worse. While I try to go on for Maxine and Daddy I feel that it won't be long until I meet him. I can't understand why it had to be Jack. We loved him so and we were so close to each other. We just lived for each other. His lieutenant wrote me stating that Jack had been hit by the Japs, he was in the lead and after he fell he lived long enough to raise up, signal his men and saved the lives of 35 men. When they saw him raise up they finished him off with machine gun fire.

We were always so proud of him. It wasn't just because he died a hero, but if I did raise him, he was always a wonderful boy, never causing us a minute of worry until after he got in service. He started carrying papers when he was 10 years of age. Buying his own books and sending himself through school.

Here it is Monday morning and I have always done my work and run over the house in general routine, but now here it is Monday morning and our washing isn't hanging on the line. I never would have believed that I could have changed so. Daddy seems to think that this was his first combat but I feel that he has been in action for some time and his whole four years hasn't been any Sunday School Picnic. I feel bitter towards the services for they never let him come home once, and he joined in 40 and they are eligible for 30 days a year in peace time. However it is done for some purpose beyond our control, but there are times I feel I never can stand it.

It's the horror of war. Suffering humanity, death and destruction. Jack was pretty young to be shot down in cold blood. I as well as thousands of other mothers did not raise our sons for targets. I am not one bit patriotic when it comes to giving him up. He would have been 23 had he lived until September 10th. You spoke about him not writing to you. He was miles away from your place and must have been killed immediately after he left. Wish we could have the pleasure of meeting you, for my friends and Jack's friends were always friends of ours.

I want to thank you and your family for the pleasant time you gave our son during his stay in Melbourne. While he had never written to us telling us about his visits to your home, but as you know our letters were scarce because he refused to write. But he would always say, 'Please mum, don't worry about me so much. I'll be back. I promise you.' Well may God Bless you and your family Mrs Darke and I would love hearing from you again.

Sincerely yours

Mrs Williams

Mrs Williams corresponded with my mother until about 1950 after which the letters dried up as she tried to move on with her life. Some of her last communication arrived in 1949 sharing details of Jack's formal burial at Greenwood Cemetery in his home town of Canton. The local newspaper covered the service and its headline recorded.

'MARINE HERO JACK WILLIAMS RITES SATURDAY:
Corporal Gives Life to Save Others in Spearheading Attack. Jack Williams was buried with full military honours. His medals, The Purple Heart, Combat Action Ribbon, The Pacific Campaign Medal with Two Battle Stars, The Marine Corps Good conduct Medal and The WW11 Victory Medal adorned his flag draped casket. The American Legion provided a burial party, firing squad, colour guard and pallbearers. At the conclusion of the ceremony the last post echoed around the cemetery and the American flag was presented to his mother.'

In what was to be the last letter, Mrs Williams wrote to my mother that the ceremony was a dreadful ordeal for the family to endure, but that they all felt so much better knowing that Jack was on American soil. She added,

> I wish you folks could have been here, for his service, if there is such a thing as a pretty funeral, Jack's was the most beautiful I have ever seen. It was largely attended and so many beautiful flowers. And we were so proud to know that he was held in such high esteem in the community in which he had lived. We are trying so hard to pick up the pieces of our lives as he would want us to do.

Clearly the formal return of their son's remains provided the Williams family some comfort in the excruciating loss of their beloved son. Over the remaining years their visits were constant on birthdays and of course the anniversary of his death. In spite of Mrs Williams initial belief that she would join Jack soon after his death, she lived until 1985, her husband having pre-deceased her in 1970. If you happen to walk through the Greenwood Cemetery you can see that Gus and Rachel Williams rest with Jack for eternity. Here in Australia, on the other side of the world, my mother also carried Jack Williams in her heart until her own final days. Liberty Hall had given its family and their friends a fleeting taste of freedom and joy before grief and sorrow called for last drinks.

THE BLOODING

Men from all walks of life were forced to confront the horrors of war. This is the true story of my father-in-law and the seminal event of his World War 11 experience.

As he embraced his new bride on the steps of the Melbourne Grammar Chapel, Alan Rosenhain knew that he had been thrice blessed. He had wooed and won the hand of the vibrant and beautiful Joan Storey, he had graduated with a prestigious medical degree and had, within the last week, been appointed a Captain in the Australian Army Medical Corps. Standing amid the well wishers, his future shone as brightly as the three pips on his shoulders. His current euphoria was only tempered by the news that the authorities had directed him to report to his posting in three days time. On his wedding night, Alan braced himself for tears when he broke the news about the curtailed honey moon plans to his adoring wife. The honeymoon which he had long anticipated, was like all war time nuptials, to be cut cruelly short.

Mindful of his minimal practical experience, the authorities instructed Captain Rosenhain to report to an army hygiene training camp in Sydney before being deployed to Cowra; a country town some 310 kilometres south west of Sydney and 190 kilometres north west of Canberra. With a population of three thousand, its sole claim for military attention was determined by the fact that it housed a camp for prisoners of war and other internees some three kilometres beyond the town. The prisoners were housed in a dodecagon while three Australian Military Forces compounds were placed on the outside perimeter. The majority of the inmates consisted of over 1500 Japanese prisoners who had been captured in the Pacific Islands, while a scattering of other nationalities included Italians who were residing in Australia at the outbreak of hostilities and some Koreans. The sullen Japanese prisoners were a marked contrast to the jovial Italians whose unthreatening dispositions saw them despatched to neighbouring farms each day, where they milked cows and spread smiles like sunshine among the families they served. With soldiers and machine guns to guard those within the wire enclosure, the posting promised to be a gentle inauguration into both the army and medicine for a man of Captain Alan Rosenhain's refined sensitivities and sheltered family life.

As the eldest of three sons to entrepreneur Kurt Rosenhain and his wife Agnes, Alan had been denied nothing. While countless families faced blighted futures through the loss of their sons in the Great War, Kurt had established the foundation of his empire to sufficiently satisfy the expectations of Agnes Church and the demands of a growing family. Although his father was constantly introducing the boys to new gimmicks and gadgets that he imported, Alan had little interest in such wizardry. He was a thinker and dreamer whose gentle disposition was refined by the music and literature that flooded the expansive Brighton home. Beethovan's Fifth Symphony aroused a certain ecstasy in him that was best enjoyed on a summer afternoon as he lazed on the manicured lawns of the extensive gardens. If not reading or listening to the music of the great composers, he could rouse himself to a few games of tennis on the adjacent grass court. His two brothers were positively ruthless in such impromptu tournaments, but Alan rarely worked up a sweat as he placed meticulously angled volleys that out manoeuvred his brothers' brute force. With the shimmering sands of the beach merely within walking distance, his was a halcyon childhood that knew neither anger nor angst.

Cocooned in wealth and refined sensibility, Alan remained an often solitary figure as he embarked on his formal education at the elite Melbourne Grammar School. Here he found the boys coarse and the elderly post war masters brutal and uncompromising. Eschewing the muddy brawls of Australian Rules football, Alan was more than happy to don his immaculate whites to prove himself a varied and delicate stroke player in the school cricket team. His tanned face and arms contrasted with his crisp regalia and his warm smile ensured that he cut quite a handsome figure to the admiring spectators. While he considered himself fortunate to have wedded Joan Storey, she had not exactly made a bad choice either.

Although Alan delighted in his officer status after his arrival at Cowra, he did not relish the routines and regulations of the army. Life was lived on the double and at full volume. He longed for the aesthetics of his former life and the resumption of the nuptial delights that marriage had now promised. Although he found the company of his fellow officers a tad boorish, he was not critical of them. He knew his own measure and accepted that he was not really a man's man and understood that he would find little joy in the boisterous antics and risqué jokes of his comrades. At the completion of his rather basic daily clinic, which extended to coughs and blisters, he preferred to retire to his quarters, rather than scoff beers amid the bluster of the bar. In such privacy he could read, listen to music or write passionately to his beloved Joan about the delights they would enjoy when circumstances prevailed. As he read and reflected each night, he felt grateful for the opportunities that life had delivered to him. Unlike so many men who had served over the past four years, he would never have to experience the mud, blood and guts of actual combat. Such servicemen he thought, may well return with medals and stories (if they returned at all) but all things considered, he was more than happy with his wife and his current sinecure.

Clearly neither Captain Rosenhain nor anyone else had any idea of the brutality that was brewing within the compound. For the terms of their incarceration, the Japanese prisoners of war had rarely declared their proper names. Most accepted that it was better for their families and loved ones back at home to believe them dead rather than be held as prisoners. They had all been indoctrinated and imbued with the Japanese military creed of *Bushido* that held that death at your own hands was preferable to capture. For months they had held furtive meetings to address their shameful circumstances. They planned to construct an honourable end for themselves and in so doing, kill as many of the enemy as possible. None of them had an iota of knowledge as to where they were imprisoned, or what existed in the vast landscape beyond the perimeter fence. They had observed on their transfer from Sydney, the featureless, barren paddocks that were a far cry from the lush, orderly rice paddies and settlements of their homeland. They knew that this harsh land offered no sanctuary. They did not care, their honour was all that mattered.

For months the Japanese prisoners meticulously planned and prepared for their escape. Under cover of night in their bare wooden huts, they huddled in their plum dyed prison uniform to secretly sharpen mess knives, chisels and garden stakes that they had purloined. Additionally, they had collected stray nails and pieces of metal to hammer into the heads of the baseball bats that they had acquired for recreation. All were silent and lethal weapons. Over countless nights the prisoners had secretly scrutinised the rather lackadaisical and complacent guards mounting the early morning watch. It was easy to recognise these as the hours of vulnerability. By 4 August 1944 their resolve was clear. The crisp, cold nights heightened the guards' inattention as they shuffled under their great coats to sustain warmth. Tomorrow morning would be the time to act.

That night the prisoners gathered to make their final farewells and sharpen their desire with toasts of illicit alcohol. Dressing themselves as warmly as possible and carrying blankets from their beds to buffer the barbed wire in their path, they waited until 0150 hours to set alight their huts and sound a rallying cry with an old battered bugle. Thus galvanised, the Japanese prisoners stormed the perimeter fences in an attempt to gain control of the machine gun posts to either kill the Australians or themselves. With huts ablaze, their bloodthirsty shrieks and yells rent the bitter night air as they charged like a red wave. The four shocked and complacent sentries manning the machine gun posts were quickly overpowered by hordes of manic escapees. Nonetheless, two privates managed to reach the Vickers gun before the Japanese and quickly began mowing down the waves of escapees. Inevitably the sentries were overpowered and brutally bludgeoned and stabbed to death. Then with little immediate opposition, the prisoners threw their blankets over the remaining barbed wire or jumped from the towers to facilitate their escape. They cared nothing about death, but dishonour was unthinkable.

As the alarm sounded and reserve troops were summoned, Captain Alan Rosenhain was hurriedly roused and commanded to proceed directly to the Cowra Base Hospital to prepare the operating theatres for anticipated surgery. Stunned and shocked by such a rude awakening, Alan's confidence wavered. His surgical experience was minimal. His exposure to traumatic wounds was virtually non-existent. How would he cope with the dire circumstances that confronted him. Realising that he now had to put to the test the seven years of medical training he had received, he was determined to remain calm. Challenging this resolve was the rumour that the escapees were headed for Cowra and as his wife was lodging there, an all pervasive panic consumed him. Again he was blessed with good fortune. Elva Trengrove from the AWAS training camp was ordered to convey the bodies of the slain guards, Privates Jones and Hardy to the Cowra Hospital and morgue. Captain Alan Rosenhain was to accompany her and by virtue of that chance meeting, the raw young doctor was able to draw on her strength and compose himself for the task ahead.

Back at the camp, the officers and reinforcements surveyed the carnage of burnt bodies and corpses impaled upon the three rows of barbed wire. While most prisoners had chosen to escape, several had hanged themselves in the burning compound. The violence and fervour of the escapees' intent horrified the Australians and in fear of the threat to the local population, hundreds of soldiers were despatched to scour the immediate neighbourhood and warn the locals. Many prisoners were quickly rounded up, begging their captors to shoot them. Apart from one situation, such exhortations were not considered. By 2.00p.m. that day, August 5 1944, Captain Alan Rosenhain was operating on wounded Japanese escapees. The raw young doctor had never seen anything like the injuries he witnessed that day. The deceased men admitted to the hospital had suffered horrific wounds. Some had thrown themselves under advancing trains, others had hanged themselves, while burns, stab wounds, gaping machine gun caverns and blunt trauma injuries accounted for further mortality. In his training, Captain Alan Rosenhain had witnessed and certified death, however he had never seen death on such a large and gruesome scale. Of the 334 prisoners who escaped, 231 died and 107 received severe and brutal wounds. Four Australian servicemen were killed and Captain Rosenhain was ordered to perform autopsies on them in order to furnish a Court of Enquiry as to the causes of death. It was not an easy or pleasant task. As his saw and scalpel dissected the corpses of his comrades, his stomach heaved but he gritted his teeth. He knew that responsibility demanded high expectation.

As August 10 dawned, the last of the escapees had been recaptured. For days they had wandered around the sparse and unfamiliar terrain, not knowing or caring what to do. Once recaptured, the wounded were brought to the hospital for treatment. None of the injured acknowledged any degree of pain, as their bones were set or their slashed and gaping wounds were stitched. Silently, Dr Rosenhain set about his grim duty, tempering his personal rage at the brutality of his patients with the dictates of the Hypocratic Oath. The prisoners expressed no gratitude, but their body language reflected despair. They had survived. They had failed. When he left the hospital that evening Captain Rosenhain felt like Macbeth, that his hands were steeped in blood. The blood of madmen who had ruthlessly murdered Australian soldiers in their quest to die honourably. From his limited experience he felt there was very little to be revered in violent sacrifice. Blood, bone, tissue and shit were, the universal remnants of a human life when we all meet our end. Honour was in life itself, the beauty of how we savoured our existence to enhance the future. Bloodied, exhausted and bewildered, Alan could not fathom a credo that valued death over life. It was the very antithesis of all he had been taught to venerate. As he returned to his quarters and fell upon his bed, he began to realise that he had borne witness to the biggest prison escape in Australia's brief history. However, the Cowra Breakout was a huge blot on Australia's security and the army's efficiency. War time censorship prevailed and his experiences were to be strictly classified. His silence was expected and demanded.

As Captain Rosenhain lay on his narrow bed and processed the dramatic carnage of the past five days, his reverie recalled a crowded class room in his final days at Melbourne Grammar School. The boys were to sit their matriculation exams in the following week. The crusty, old physics teacher, perhaps as a gesture of farewell, was circulating the room asking each boy what course he wished to pursue at the University of Melbourne. He was a hopeless teacher and Alan loathed him for his ineptitude. Yet like the bully he was, he eyeballed the students and poked a bony finger at them as his interrogation probed the aspirations of each boy. To some he shrugged his shoulders in despair, to others he raised his eyes to the heavens in a cruel suggestion that they would need divine intervention. There was nothing affirming about the man. Alan shuddered as the decrepit old fraud moved toward him. Even after all of those years, he could still remember the teacher's acerbic tone as he questioned,

'And what about you Rosenhain? What do you want to study?'

'Medicine Sir,' had been Alan's usual quiet response.

'Medicine Rosenhain! A dilettante like you. You'd never make a doctor,' he mocked derisively. 'Too soft!'

Alan felt his gorge rising as he recalled the ignominy of that long distant occasion and how he had bitten the side of his cheek to ensure that he didn't respond.

'You miserable old bastard,' he ruminated as he gritted his teeth. 'It's a pity that censorship and security prohibits me from shoving this experience right up your fundamental orifice when I get back. After what I've seen and done these past five days, I can cope with anything. I'll be a better Doctor than you ever were as a teacher!'

It was an oath that he fulfilled. For upon return to civilian life he blended his training and his baptism at Corowa to honour life and service until he relinquished his registration to practice, aged ninety one!

NEVER FORGOTTEN

Most men never returned from war as the same person who left. Society also changed rapidly, which made adjustment difficult after their demobilisation. In this true story about my father after his service in World War II, his love and sensitivity remain immutable.

When my father Jim was a boy of about thirteen, the streets of Melbourne resounded to the musical clip clop of iron hooves across cobblestone streets. Large Clydesdale draught horses with muscles bulging, heads bobbing and huge feathered feet lifting rhythmically, strained the leather harnesses to pull the brightly coloured carts and wagons around the suburbs. In those days much of the household shopping was purchased from the milkman, the baker, the iceman and the Chinese market gardener who plodded the streets to sell their wares.

Everyone knew each of the suppliers and their routines. In winter Jim and all the children in his street craved a piece of fresh bread when the warm aroma wafted from the bread cart as the baker's glossy, red van opened its doors. Shelves of block loaves and buns on paper trays, were a mouth watering sight to children who played endlessly in the streets and worked up a hunger kicking a parcel of rolled paper as a football. In summer, when the heat bounced off the rows of red brick houses into streets like cauldrons, the ice cart drew squeals of delight. Jim and his friends flocked around the driver as he opened his doors to hook a large block of ice onto a hessian sack draped over his shoulders. Invariably the driver would give in to the children's entreaties, and jabbing at the blocks with an ice pick, he would chip off splinters of ice for all to suck. These slivers were dragged over grimy, hot faces or wrapped in the corners of handkerchiefs to suck before they melted.

While the baker and the ice man were seasonal favourites, the little Chinese market gardener in his black and yellow cart provided real colour.

The reds, greens and oranges of his fruit and vegetables always looked enticing as he hung his scales from the roof of the wagon to weigh the goods, wrap them in newspaper before negotiating a price and dispensing change. He came twice a week, usually on a Tuesday and Friday, the latter to provision people for potential weekend family gatherings. The most regular of all though was the least seen and the most important. This was the milkman . He was the neighbourhood alarm clock, as the slow clop of hooves and the steady squeak of wheels, echoed against the clink of glass milk bottles being placed on the front doorstep of each house.

Jim loved the parade of carts selling their wares, not just for the goods they conveyed, but for the horses that worked so hard and dutifully for long hours each day. His father owned a dairy in Balaclava. As a small boy Jim had watched his father un harness the horse at the end of his round, brush and feed her, before finally polishing the tack. Jim often helped when he came home from school. He grew to learn the complex pieces of harness from the bit and bridle through to the collar, the belly band and the breech straps. Occasionally he would earn a penny from his father for rubbing down the mighty draughthorse at the end of the day. He loved the silky texture of the mane and the big, fluffy feet as he gently drew the curry comb through the lengths of fur. The horse often snorted and stamped its hooves if the comb caught in the tangles, but Jim was not daunted or intimidated. He simply talked soothingly and gently as he combed and watered the huge beast.

Hitching up the horse and wagon each morning and loading it with milk and cream bottles had been his father's routine for years. Darke's Dairy was known throughout the suburb. Jim felt enormous pride in his father's reputation for both the integrity of his service and the gentleness of his great draughthorse, which knew its round so well that she would wait for Arthur at the end of each street. Sadly though, the routine was to end. The Great Depression of the 1920's hit the dairy like a brick wall. The daily necessity of milk grew beyond the means of many. Mothers bought less milk and watered it down, or would knock on the door with an empty billy can and plead with my grandfather for, 'just a little drop if you could spare it Mr Darke.'

The reality was that Jim's father, Arthur, could not spare it, but he gave each hollow eyed mother with a small child balanced on her hip a generous scoop from the churn, as he could not bear to turn them away empty handed. Such generosity was incompatible with running a business and gradually the dairy went broke and my grandfather had to join the ranks of the unemployed himself. Times were cruel and so Jim, who, in spite of being a very able student, was forced to leave school in the hope that he could find work and help support the family. As a boy of only thirteen, his skills were limited, but one day he heard a whisper when buying a loaf of bread, that there might be a vacancy at the bakery in South Yarra.

Without mentioning it to his parents Jim walked from St Kilda to the South Yarra bakery where he enquired about the job. The manager's instinctive response was to laugh at the slender teenager, exclaiming that the required hours and heavy duties were 'men's work,' Undeterred, Jim pleaded with him and assured him of his competence with draughthorses and his ability to work the required long hours. Given that a boy's wage was less than a man's, I suppose that the manager figured that it was worth giving the lad a try, and so after being kitted out with a uniform, Jim arranged to begin work the following Monday.

Jim was awake before his alarm sounded at 3.00 a.m. to herald the start of his working life. He was excited at the prospect of driving a baker's cart and the round given to him was familiar St Kilda territory. Nonetheless, he was all fingers and thumbs as he slid the bridle around the huge draughthorse to bring her in from the paddock to the harnessing yard. He had been allocated Bonnie, a brown and white Clydesdale who stood at 18 hands. Although she was only five years old, she knew the routine and waited patiently while Jim mixed a third of bran to two thirds of oats for her breakfast. While she was eating, Jim laid out her tack and then allowed the great animal to drink long and deeply before he began harnessing her. Gently, but authoritatively, he placed the bridle over her head, positioning the noseband and blinkers to allow the bit to fit comfortably into her mouth. Talking gently and stroking her withers he then moved into the familiar rhythms of placing on the collar, the saddle housings and breechings before backing her into the shafts. When finished he stood back in wonder. She was beautiful. Slowly, Jim moved to stroke her velvet nose. Instinctively, she nuzzled her head under his arm to flush out a carrot he had hidden for later.

A partnership had begun. For eleven years Jim and Bonnie followed the same routine. He learned to whistle to her in the grazing paddock of a morning and she would prick up her ears and walk over to the harnessing shed. A carrot or apple was her morning entrée after which she was properly fed and harnessed. Whilst she stood patiently, Jim would fill the trays in the little red cart with 'Stockdale's Bread' emblazoned on the sides. Usually he carried up to 460 loaves and an assortment of buns that would be distributed to around 350 customers a day. The round he and Bonnie undertook covered the run down cottages in Inkerman Road up to the rich Italianate Mansion at Ripponlea. At each house Jim dismounted from the little driver's seat, filled his wicker basket with orders and walked briskly from door to door. Most women left an empty tin with the order and the money inside it on the front doorstep. Quickly, Jim filled the order and placed the change in the tin and then gave Bonnie a whistle to walk on. Bonnie soon knew how far to go before stopping at the appropriate house or corner, and waited for her driver to climb up and direct her into another street.

The round was usually completed by 2.00 p.m. and although weary, Jim knew from experience that the day was far from over. When all his muscles ached and his eyes began to close with the rhythm of the gentle clip clop back to the bakery, he knew that the most important task awaited. Bonnie had to be unharnessed, watered, fed and groomed. With leaden arms, Jim stowed the tack, blackened Bonnie's hooves with boot polish and then brushed her until she shone. He chatted to her as he groomed, praising her for her good work and gentle temperament. In response, she would snort in his hair and give a gentle nuzzle in the direction of his pocket for her carrot or apple, before being released back into the paddock for the night. Jim rarely arrived home before tea time and after that, the rigours of the day on a young teenage body ensured that sleep was never long in coming.

As months turned into years Jim and Bonnie became a well known team. Bonnie knew Jim's smell, his walk, his kindness, but most of all his whistle and the treat that followed. Jim continued to care for Bonnie lovingly and her maturity saw her coat glistening and her flanks filling out. It was the same for Jim. The long hours, exercise and fresh air in all weathers, saw him grow tall and develop strength and endurance. He was a handsome man who took each day on its merits. He took pride in his horse and his work however humble, and more importantly his labours from adolescence to manhood, had ensured that his family had survived the Depression.

Perhaps Jim and Bonnie would have continued their work until the inevitable replacement of horses with cars but instead the war intervened and Jim left the bakery to join the army. When Britain first entered the war against Germany, Jim happily joined the 'Chokkos' otherwise known as the Army Reserve. Australia's losses of over 60 000 men in the First World War deterred him from volunteering to fight the Empire's battles in Europe. He loved Australia and was sufficiently well read to see the potential threat to his homeland from the Japanese. As soon as this threat became a reality, he immediately enlisted. He proudly served with the AIF for four years as a signalman in Darwin during the Japanese attacks. He relished army life and the easy camaraderie among the men. He loved the open spaces of the Northern Territory and blazing night skies that a city upbringing had denied him. Thriving on the responsibility of army life he was quickly promoted to sergeant and it was suggested that at the end of the war that he sign up for Officer's School and become a career soldier. However, like thousands of others over this period, Jim married and had a daughter and saw little of his small family until the war's end. His wife, Joyce, was not enthusiastic about the possibly of the peripatetic life as an army wife and so Jim was demobilised in 1945. Now, with his own family to provide for, he was yet again looking for employment.

It was no surprise that Jim returned to the bakery to see if his record of previous exemplary service would open doors to another job. Much had changed though in the intervening war years: new management, and of course the motorisation of the delivery system. Jim was deeply disturbed by the latter change as a part of his desire to return was to see Bonnie again. What had happened to her? Had she died? Jim needed to know. The new management didn't know and didn't care. War had changed the world's priorities. Life now revolved around efficiency and expediency and the days of the cart horse were long gone. Jim left the bakery that day feeling despondent and a bit disillusioned. He valued integrity and decency and felt that his faithful companion had been abandoned and who knows, perhaps she felt abandoned by him. That thought did not sit well with Jim and so he began to make enquiries.

Not surprisingly his quest was furthered in a chance meeting in a pub. He bumped into an old Stockdales' workmate who had stayed on with the firm during the war as he was deemed unfit for active service. He reminisced that he was still there when the Clydesdales were gradually phased out and was as sad as Jim to see the end of an era. Naturally Jim asked what had happened to the horses and learned that they had been sold to a farmer in Drouin, a Gippsland farming community. Inspired by this snippet of information, he wrote numerous letters to various dignitaries, including councillors and the Mayor of Drouin, soliciting any knowledge of the horses. Weeks passed and then unexpectedly, he received a reply from a councillor. He remembered that he had a mate on a farm who had bought some of the horses. An address was included.

After contact with the farmer, but no knowledge of which horses the man had bought, Jim, boarded a train to Drouin. It was about a three hour trip in those days and when he arrived he had to seek directions to the farm address he was given. Jim was always a great walker and so the five mile hike to the property never bothered him. It was a pleasant day and he was on a mission. After about an hour and a half he reached his destination and made his way up the long driveway to speak to the farmer. As he strode up the hill he glimpsed a form under some trees in the distance. The paddock was adjacent to the driveway and he was sure that the shape was that of a Clydesdale.

The sun was bright and low in the sky as he squinted and tried to focus through its setting rays. The horse seemed brown and white, but then again there were a number like that in the bakery mob. Jim stood. He stared and then with a measure of uncertainty, he whistled. He whistled with the same inflection that he had trilled for eleven years. Did he detect a movement? He whistled again. He thought he detected a shake of the head and mane and then, slowly the great horse moved out of the shadows. Excited, he gave another whistle and this time called her name too.

Now there was no hesitation. The great beast broke into a canter and as its giant, feathery, hooves thudded across the paddock towards him, Jim knew that the magnificent creature was indeed his Bonnie. He kept calling her name again as she neared to the fence rail and then she stopped and looked at him. Her coat had lost some of its lustre, but her big, brown eyes were bright and she was still firm around the flanks. She had worked hard and long at the bakery, but it was clear to Jim, that whoever had bought her, demanded nothing, except that she enjoy the pasture that she so richly deserved. As Bonnie edged towards the rail, Jim whispered to her and held out his hand to stroke her face and neck. Instinctively and knowingly, she stepped forward and nuzzled under his outstretched arm. With gentle sniffs she pushed her way down Jim's side into his pocket. Then, with a loud snort, she extracted the waiting apple and leaned into Jim's shoulder to savour the delight. My father's tears dripped on Bonnie's neck and ran down her leathery nose as they held that embrace. Instinctively, my father knew all was well. For all the battles both he and Bonnie had fought this was the sweetest of victories.

ABOUT THE AUTHOR

Carol Rosenhain is a Melbourne writer and historian. Educated at Monash University, she has been a passionate teacher of English and History at senior secondary level for many years. In her professional capacity, Carol has been Head of the English Faculty at Genazzano FCJ College and a State Examiner of VCE English. She has presented seminars at the Victorian History Teachers' conferences and has co-authored with Jill Fenwick *'South Africa: From Settlement to Self Determination'* Carol has also written commercially for artists and galleries including *'The Art of Paul Margocsy; A Biography'* Her most recent publications; *'The Man Who Carried the Nation's Grief,'* and *'The Men Behind the Myth,'* both explore fresh perspectives on selfless individuals and families who served in the Great War.

www.ingramcontent.com/pod-product-compliance
Lightning Source LLC
Chambersburg PA
CBHW022119040426
42450CB00006B/758